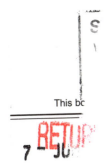
Don Gres

Roofing Failures

Roofing Failures

Carl G. Cash

Spon Press
Taylor & Francis Group

LONDON AND NEW YORK

First published 2003
by Spon Press
11 New Fetter Lane, London EC4P 4EE

Simultaneously published in the USA and Canada
by Spon Press
29 West 35th Street, New York, NY 10001

Spon Press is an imprint of the Taylor & Francis Group

© 2003 Carl G. Cash

Typeset in Sabon by
Integra Software Services Pvt. Ltd, Pondicherry, India
Printed and bound in Great Britain by
MPG Books Ltd, Bodmin

British Library Cataloguing in Publication Data
A catalogue record for this book is available from the
British Library

Library of Congress Cataloging in Publication Data
Cash, Carl G.
  Roofing failures / Carl G. Cash.
    p. cm.
  Includes index.
  ISBN 0–415–29925–X
  I. Title.
TH2391.C37 2003
695—dc21
                                2002155094

For Liz, my wife –
written during the 50th year of our marriage.
For our wonderful children, their spouses,
and our grandchildren.

Contents

List of tables   x
List of figures   xi
List of details   xii
List of haikus   xiii
List of abbreviations   xiv
Conversion factors   xv
Preface   xvii

PART 1
Introduction   1

1 A snapshot of the roofing industry in 2001   5

2 Low-sloped roofing systems and materials   10

3 Low-sloped roofing systems and materials (continued)   27

4 Structural decks and thermal insulation   38

5 Steep-sloped roofing systems   52

6 Flashing   66

7 What is failure?   84

8 Performance vs prescriptive specifications   94

PART 2
Case studies   99

9   The case of the leaking book warehouse   101

10 The case of the shattered slates   105

11 The case of the department store's splitting return 107

12 The case of the propriety products 109

13 The case of the splitting membrane 111

14 The case of the unacceptable design advice 113

15 The case of the skilled maintenance man 115

16 The case of the roof over the rare paper storage 117

17 The case of the wide spread flame 119

18 The case of the dissolving shakes 122

19 The case of the leaking gymnasium 124

20 The case of the flapping glass fiber felts 126

21 The case of the leaking condominiums 128

22 The case of the corroding foam 130

23 The case of the heavy glass fiber shingles 132

24 The case of the moving insulation 134

25 The case of the phantom deck movement 136

26 The case of the lightweight insulating concrete 138

27 The case of the shrinking insulation 140

28 The case of the gooey felts 142

29 The case of the ice castles 144

30 The case of the tile 146

31 The case of the skaters' cracks 148

32 The case of the phased felt plies 150

33 The cases of the fastener backouts 152

34 The case of the blistered shingles 154

35 The case of "1+1=4" 157

36 The case of the cold process roofing 159

37 The case of pressure sensitive adhesion 161

38 The case of the fire retardant plywood 163

39 The case of asphalt dispersion 165

40 The case of the liquid applied waterproofing 167

41 The case of the blistered airport roof 169

42 The case of pesky sea gulls 171

43 The case of the distant expansion joints 173

44 The case of the ill wind 176

45 The case of the improper waterproofing 178

46 The case of the poorly vented roof 180

47 The case of the missing facer adhesion 182

48 The case of the noisy roof 184

49 The case of the severe hail storm 186

50 What have we learned? 188

Appendix A – Roofing trade names 191
Appendix B – Roofing industry websites 235
Appendix C – Answers 239
Appendix D – Bibliography 246
Index 249

Tables

1.1	Roofing sales by application, 2001	6
1.2	Type of low-sloped roofing sales, percentages, 2001	7
1.3	Type of low-sloped roof insulation installations, percentages	7
1.4	Types of steep-sloped roofing sales, percentages, 2001	8
1.5a	European roofing market, millions of square metres	8
1.5b	European roofing market, thousands of roofing squares	8
2.1	Average and minimum years of service life, low-sloped roofing systems	11
2.2	Mean years of durability at various thermal loadings	12
2.3	Mean percent surviving and life cycle costs	13
2.4	Typical physical properties of unexposed low-sloped roofing membranes	16
2.5	Roofing material mass and equilibrium moisture content	18
2.6	Roof drainage guide	20
4.1	Maximum steel deck panel spans recommended by Factory Mutual Research Co.	40
4.2	Physical properties of low-sloped roofing thermal insulation	46
4.3	Thermal properties of some typical building materials	49
5.1	Steep-sloped roofing mean life, fire classification and life cycle cost	53
5.2	Typical composition and physical properties of asphalt shingles	55
5.3	Fire test interpretations; Classes A, B, and C	57
5.4	Estimated thermal expansion of some building materials	62
9.1	Laboratory report	103
34.1	Dry felt mass, percent saturation, and water absorbed	154
43.1	Load–strain testing of built-up roofing samples	174

Figures

1.1 Roofing systems classification   6
2.1 Horizontal roofing surface temperature graph   11
39.1 Polymer in a sea of asphalt   165
39.2 Just after phase inversion   166
39.3 Final dispersion of asphalt in a polymer   166

Details

6.1   Roof drain   70
6.2   Vent pipe penetration (metal cap)   70
6.3   Vent pipe penetration (EPDM counterflashing)   71
6.4   Multiple pipe penetration   72
6.5   Typical curb flashing   72
6.6   Metal duct penetration   73
6.7   Equipment support blocking   73
6.8   Roof hatch flashing   74
6.9   Low parapet wall   76
6.10 Gravel stop   76
6.11 Gravel stop transverse joint   77
6.12 Expansion joint   78
6.13 Relief joint   78
6.14 Typical nailing patterns   79
6.15 Metal coping   80
6.16 Insulation and ply layout over concrete deck   81
6.17 Typical EPDM counter flashing detail   81

Haikus

First line

A performance spec.	95
A phenolic foam	48
All buildings, you know,	79
All insulation	38
Always remember:	92
Asphalt-glass shingles	86
Big cost improvements	172
Blindfolded justice	99
Buy by competence	190
Coal-tar pitch roofing	24
Curiosity	88
Designers design.	69
Detailed peer review	68
Face reglet details	67
Failures are due to	85
Foam insulation	48
For steel deck advice	38
Good roofing systems	88
Half a cap flashing	68
Holes and gravity	88
I play a cracked lyre	xviii
If stupidity	190
If you examine	86
Improve your roofing	189
Inverted roofing	68
Lightweight is fancy	41

First line

Lignin-excelsior planks	42
Manufacturers	59
Materials don't fail	91
Most roofing failures	188
New and untested	189
On seal-tab shingles	86
Or equal clauses	15
Owner's competence	114
Phony roofing slates	64
Polyiso salesmen	49
Roof decks that contain	41
Roofing is joyous	2
Roofing is simple	20
Roofing monitors	87
Stinky foam of glass	45
The final chapter!	190
To assume usually	32
Torch applied roofing	29
Twenty years experience	14
Unfilled pitch pockets	67
Unless glue hardens	30
Used like a lamp post	91
Warrantees and snow	14
Werner Gumpertz sez:	61
When you specify,	14

Abbreviations

APP	atactic polypropylene
Btu	British thermal unit
C	thermal conductance
CPE	chlorinated polyethylene
CSPE	chlorosulfanated polyethylene – Hypalon
ECB	ethylene copolymer bitumen
EMC	equilibrium moisture content
EPDM	ethylene-propylene-diene terpolymer
EVA	ethylene vinyl acetate
ft	foot
in.	inch
J	Joule
k	thermal conductivity
kN/m	kilonewtons per metre
m	metre
mm	millimetre
OSB	oriented strand board
PMA	polymer modified asphalt
PMC	polymer modified coal-tar
PUF	polyurethane foam
PVC	poly [vinyl chloride]
R	thermal resistance ($R = 1/C$)
SBS	styrene-butadiene-styrene block copolymer
SEBS	styrene-ethylene-butadiene-styrene block copolymer
SIS	styrene-isoprene-styrene
TPO	thermoplastic poly olefin

Conversion factors

	SI units	Multiply SI units units by	To find inch-pound units
Application rate	litre per square metre	2.044	UK gallons per square
Application rate	litre per square metre	2.454	US gallons per square
Area	square metre	0.108	square (100 square feet)
Area	square metre	10.764	square foot
Area	square millimetre	0.00155	square inch
Breaking strength	kilonewton per metre width	5.714	pounds per inch width
Density	kilograms per cubic metre	0.062	pounds per cubic foot
Impact	joule	0.738	foot pounds
Length, width, thickness	metre	3.281	foot
Length, width, thickness	millimetre	0.0394	inch
Length, width, thickness	micrometre	0.0394	mil
Mass	gram	0.035	ounce
Mass	kilogram	2.205	pound
Mass per unit area	grams per square metre	0.0295	ounces per square yard
Mass per unit area	kilograms per square metre	20.483	pounds per square
Mass per unit area	grams per square metre	0.0205	pounds per square foot
Pressure	pascals (N/m^2)	0.0209	pounds per square foot

Pressure	kilopascal	0.145	pounds per square inch
Slope	percent	0.12	inches per foot
Speed	metres per second	2.237	miles per hour
Speed	kilometres per hour	0.621	miles per hour
Speed	metres per second	2362.2	inches per minute
Thermal conductance, C	watt per square metre · kelvin	0.176	Btu per hour · square foot · °F
Thermal conductivity, k	watt per metre · kelvin	6.935	Btu · inches per hour · ft^2 · °F
Viscosity, dynamic	pascal seconds	0.1	poise
Viscosity, kinematic	square metres per hour	277.8	centistokes
Volume	cubic metre	35.311	cubic foot
Volume	cubic metre	1.308	cubic yard
Volume	cubic metre	220	UK gallon
Volume	cubic metre	264.2	US gallon

Preface

In 1986, the International Council for Building Research Studies and Documentation [CIB] formed a Building Pathology task group to study failures in buildings, to report on methods to reduce the failure rate, and to prevent repeating these undesirable results. The membership list of the task group is geographically broad, including members from 20 countries, on every continent. The task group meets once or twice a year; it trades information on industry trends. To date, their efforts to document less than ideal performances of building systems and materials has not been very effective due to resistance by many entities, including insurance companies, manufacturers, and our own inertia.

The building pathologist examines the flesh and bones (the skin and structure) of buildings to isolate the reasons for less than satisfactory performance, and communicates by teaching, activities in trade organizations, active membership in standardization societies, and professional societies, the lessons learned, to minimize the repetition of the poor performance. (One wag suggested that we should be called: building proctologists, because we are always searching for leaks.)

This book represents my effort to fulfill part of my obligation as a building pathologist to the roofing industry and the many individuals who have helped, instructed, and encouraged me throughout my career. Among these are:

- My partners and friends at Simpson Gumpertz & Heger Inc., especially Dave Adler, Steve Condren, Arthur Davies, Werner Gumpertz, David Niles, and Tom Schwartz.
- My owner and general contractor clients, including Janus Baxa, Jeff Bliss, Doug Campbell, Fred Conogue, John Cook, Floyd Gray, John Gutman, Edward Kakas, Ken Kimbrough, Richard Kobe, Ed Landry, Charles Soelner, Robert Tanner, Richard Trant, Wayne Weaver, Richard Wilt, and Paul Zimmerman.
- My international technical friends, including Sergio Croce, Coias e Silva, Mauro Maroni, Bill Porteois, Keith Roberts, Nikolaj Tolstoy, Peter Trotman, Art van den Beukel, and "Timber" West.

- My many roofing contractor friends, past and present, especially Herb Fishman, Charlie Griffiths, Stuart Grodd, Burt Karp, Mel Kruger, Bill Kugler, Bob Linck, Paul Morris, Wayne Mullis, Milt Olsen, and Neal Simon.
- The many lawyers who have instructed me, and who have in turn been instructed by me, especially: Tony Abato, Richard Andriolo, Katheryn Barnhill, David Birka-White, Greg Blackburn, Mark Bridge, Daryl Brown, Susan Cole, Phil Cronin, George Fleming, D. Taylor Flowers, Ken Gilman, Peter Goetz, Mark Grantham, John Herlihy, Donald Kemple, Francis Kathcart, Robert Lee, Bill Mattes, Michael Less, Mike Meagher, Victor Meyers, John Miller, Warren Miller, Frank Nemia, Louis Pepe, Steve Phillips, Carmi Rapport, Herbert Stutman, Paul Sugar, Steven Sutton, James Young, and Jeff Youngerman.
- The large number of technical personnel, who taught me all that I know, including: George Berry, Roger Bonafont, Paul Buccellato, Walt Butterfield, Bill Cullen, Bruce Darling, Andre Desjarlais, Sid Dinwiddie, Rene Dupuis, Mike Franks, Dick Fricklas, Charlie Goldsmith, Mark Graham, Justin Henshell, Arnold Hoiberg, Miles Jacoby, Dick Janicki, Husnu Kalkanoglu, Jayant Kandy, Tim Kersey, Joe Klimas, Hesmat Laaly, Dorothy Lawrence, Bob Mathey, Bob Metz, Jim Mollenhoff, Toby Nadel, Richard Norris, Ernie Ostic, Ralph Paroli, Helene Hardy-Pierce, Don Portfolio, Charlie Pratt, Dave Richards, Walt Rossiter, George Smith, Eric Stern, Ken Sutton, Wayne Tobiasson, John Van Wagoner, Richard Wallace, Stan Warshaw, Jim Weidman, Jack West, and John Wooten.

I owe all of you a deep debt; which I hope this book starts to repay.

I play a cracked lyre
And sing in a broken voice
The song you taught me.

Part I

Introduction

This book has several parts. The first group of chapters provides an overview of the roofing industry in the United States at the turn of the millennium (2001), the principal segments of the industry, and the sales volume within the low-sloped and steep-sloped markets. The chapters explore the characteristics of the most frequently used roofing systems, waterproofing materials, decks, and thermal insulations. Additional chapters review the many definitions of "failure", and the influence of "performance" vs "proscriptive" specifications.

The reader can check his current understanding of these matters by correctly answering 80 percent of the questions found at the conclusion of each chapter. The answers are posted in Appendix C at the conclusion of the book. The reader can just scan these chapters if their experience makes them unnecessary.

Each of the chapters in the second group contains a situation or a case history that illustrates a type of roofing failure. The names of the participants and the location of the events are altered to protect both the innocent and the guilty, because these cases are not intended to laud or embarrass anyone. They are allegories intended to be illustrations of positions or situations to avoid, preventing a repetition of the failures.

Questions posed at the end of each chapter are intended to stimulate classroom discussion, and may be used as essay questions in an examination. Definitive answers to these questions are not provided in this book due to the many possible responses to these questions. Comments are made for each case study.

The final chapter makes recommendations to owners, roofing system specifiers, general contractors, specialty contractors, and roofing researchers about methods to avoid roofing problems.

Appendix A is an alphabetic listing of trade names used in the roofing industry. Appendix B lists roofing industry related websites. Appendix C, as previously mentioned, provides answers or comments relating to the questions posed after each chapter.

This book does not make the reader a roofing system designer or specifier. There are many excellent books and responsible suppliers' literature to fill that need. The intent is rather to present an overview of the subject so that the reader can have at his command the knowledge to ask intelligent questions, and avoid some of the problems we have observed over the years.

I have taken advantage of a Japanese literary mechanism, the haiku to provide aphorisms, emphasis, and perhaps some humor to our subject. A haiku contains three lines. The first and third lines have five syllables, and the second line seven syllables. As:

> *Roofing is joyous*
> *Always remember: it is*
> *On top of us all!*

Aside from CIB, mentioned in the preface, several other organizations are frequently mentioned in this work including:

- ASTM International – 100 Barr Harbor Drive, West Conshohocken, Pennsylvania 19428-2959 – www.astm.org.
- NRCA – 10255 West Higgins Road, Suite 600, Rosemont, Illinois 60018-5607 – www.nrca.net.

ASTM INTERNATIONAL

It was once called the American Society for Testing and Materials. The shortened name reflects its growing international commitment and utilization. This society is primarily made up of volunteers of interested parties who develop consensus standards for many industries. Many argue that the standards developed are too easy; they do not represent the "best" practice. The problem is that the "best" practice varies in the eyes of the beholder. Thus, the standards developed are the best that can achieve consensus from manufacturers, users, and general interest members and sometimes represent a minimum quality standard for the industry involved.

The primary committees involved in developing standards that have a relationship to roofing include:

- C16 on thermal insulation (1938),
- C24 on building seals and sealants (1959),
- D08 on roofing and waterproofing (1905), and
- E06 on performance of buildings (1946).

The number in parenthesis is the year the committee was established. The principal roofing committee will celebrate its centennial in 2005. ASTM publications that may assist the reader include: *Annual Book of ASTM Standards, Volume 04.04 on Roofing and Waterproofing* and *ASTM Standards Relating to Materials, Systems, and Testing for Roofing and Waterproofing*.

NRCA

The National Roofing Contractors Association is one of the largest, oldest, and one of the most effective trade organizations in the United States. Their massive *NRCA Roofing and Waterproofing Manual* and their *Low-Sloped Roofing Materials Guide* and *Steep-Sloped Roofing Materials Guide* should be used by everyone interested in roofing and waterproofing.

These sources and additional resources for those with a serious interest in roofing and waterproofing are listed in an appendix at the rear of this book in Appendix D.

1 A snapshot of the roofing industry in 2001

Most of the following is developed from data reported in Professional Roofer magazine (Hinojosa 2002: 24–8). It is based on a survey of roofer members of the NRCA (National Roofing Contractors Association) early in 2002. It shows a quick estimate of the roofing sales in the United States at the turn of the millennium, the way roofing systems are classified, and the volume within each classification. These data are benchmarks, and a way of evaluating the relative importance of each market segment.

The total estimated market for roofing products in 2001 was $30.18 billion. This is slightly down from the previous year, but a 4.1 percent increase expected for 2002.

Roofing systems are first cataloged into low, dual, and steep-sloped systems as shown in Figure 1.1. Low-sloped systems are intended for use over decks that slope toward drains equal to or less than 25 percent (3 in./ft). These systems are intended to be watertight. Dual slope systems are used on either low- or steep-sloped decks. Steep-slope systems are intended for use over decks that slope equal to or more than 25 percent (3 in./ft), and are water shedding rather than watertight. Indeed, Dave Adler, one of my colleagues, says that the proper slope for steep-sloped roofing is too steep to walk on comfortably. Low-sloped roofing sales represent ~63 percent of the market in 2001.

Another way of classifying the market is by the product used for roofing new construction, re-roofing, or repairs. The distribution in each of these categories is shown in Table 1.1. The statistics reported show that in 2001, the old roof was removed 71.5 percent of the time before a new low-slope roofing system was installed. Roofers report that proportion of old roofing removed was 82.9 percent before a new steep-sloped roof was installed.

Table 1.2 shows the percentage of low-sloped roofing sales for each type of system and for new or re-roofing. Ethylene-propylene-diene rubber (EPDM) dominates the national market, but there is significant variation in the local markets. EPDM is more dominant in the northeast and not dominant on the west coast of the United States.

It is disturbing to note that 16 percent of the low-sloped decks are being roofed with steep-sloped products or systems that are intended to be water shedding rather than waterproof. In mild climates with a low quantity of

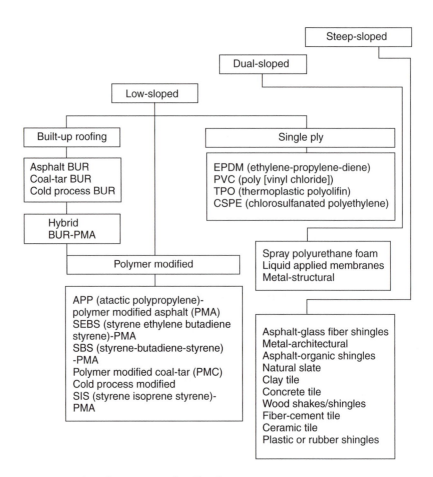

Figure 1.1 Roofing systems classification.

rain, this may be acceptable, but in most climates steep-sloped products can be effectively installed on low slopes only if they are installed over a properly installed low-sloped system. One-third of the low-slope new construction market is shared by the traditional asphalt built-up roofing and the APP (atactic polypropylene) and SBS (styrene-butadiene-styrene block copolymer) polymer modified asphalt systems.

Table 1.1 Roofing sales by application, 2001

Application	Low slope (%)	Steep slope (%)
Re-roofing	61.9	59.6
New construction	25.0	29.4
Repair/maintenance	13.2	11.0

Table 1.2 Type of low-sloped roofing sales, percentages, 2001

Material type	New construction (%)	Re-roofing (%)
EPDM	27.9	24.2
Steep-sloped systems	16.0	14.6
Asphalt BUR	12.3	15.6
SBS polymer-modified asphalt	10.6	11.2
APP polymer-modified asphalt	10.1	10.4
PVC	7.5	6.9
TPO	7.1	4.5
Structural metal	2.1	1.6
CSPE/Hypalon	1.7	1.2
Other single plies	1.4	2.6
Spray polyurethane foam	1.2	1.7
Cold process BUR	1.0	2.2
Liquid applied	0.8	1.8
Coal-tar BUR	0.3	1.5

PVC (poly [vinyl chloride]) systems had a seven and a half percent market share; the largest share by a single ply system other than EPDM. TPO (thermoplastic polyolefin) had a seven percent market share. This is quite high for a relatively new and untried roofing system.

Table 1.3 lists the insulations used in the low-sloped roofing systems, the proportion by type in 2001, and the projected proportion for 2002. Polyisocyanurate foam is in over half of the installations. Wood fiberboard and perlite are the next most frequently used. I expect that the wood fiberboard and perlite installations will increase, as it becomes evident that none of the roofing membranes should be directly adhered to any foam insulation without a cover board of some type. This will be discussed in greater detail in Chapters 2 and 3, which deal with low-sloped roofing systems.

Table 1.4 lists the percentage of roofing sales for steep-sloped roofing products. Asphalt-glass fiber shingles have the dominant market share with

Table 1.3 Type of low-sloped roof insulation installations, percentages

Insulation type	2001 (%)	Projected 2002 (%)
Polyisocyanurate	55.9	54.9
Wood fiberboard	13.1	13.2
Perlite	11.8	12.0
EPS (expanded polystyrene foam)	9.7	10.2
Composite	2.2	2.4
Glass fiber	2.1	2.5
Other	1.7	1.9
XPS (extruded polystyrene foam)	1.4	1.6
Cellular fiber	1.4	0.7
Mineral fiber	0.7	0.7

Table 1.4 Types of steep-sloped roofing sales, percentages, 2001

Material type	New construction (%)	Re-roofing (%)
Asphalt-glass fiber shingles	44.9	50.4
Metal-architectural	20.0	14.1
Low-slope products	11.9	13.3
Asphalt-organic felt shingles	4.8	5.1
Natural slate	3.8	3.0
Caly tile	3.4	2.0
Concrete tile	3.3	2.9
Wood shingles/shakes	2.9	2.8
Spray polyurethane foam	1.1	1.1
Liquid-applied	0.1	0.9
Fiber-cement tile	0.1	0.2
Metal-structural	3.7	4.2

44.9 percent for new construction and 50.4 percent for re-roofing. The next largest market share (20 percent) is for architectural metal roofing, such as standing seam or other field fabricated metal systems. This is a remarkable increase of 2.9 percent for new construction and 4.5 percent for re-roofing.

Low-sloped products were used on steep-sloped decks for a 16.8 percent market share. This is much less disturbing than the percentage of steep-sloped products used on low-sloped decks because they can be watertight if properly installed. Tables 1.5a (metric) and 1.5b (inch-pound) present data

Table 1.5a European roofing market, millions of square metres

Year	SBS	APP	PVC	BUR	EPDM	TPO	EVA	ECB	PIB	CPE	CSPE
1998	134.1	88.29	38.92	25.55	8.25	5.33	3.89	4.39	2.93	0.97	0.55
2000	134.4	95.52	40.76	21.75	8.32	6.17	3.83	3.88	2.53	0.82	0.36
2001	131.6	97.69	43.06	19.74	8.48	6.72	4.01	3.89	2.56	0.82	0.32
2002[a]	132.1	98.20	44.58	18.19	8.79	7.30	4.20	3.95	2.80	0.92	0.29
2004[a]	135.8	97.44	47.55	15.79	9.35	8.57	4.58	4.02	2.61	0.81	0.26

Note
a Estimate, abstracted from industry reports.

Table 1.5b European roofing market, thousands of roofing squares

Year	SBS	APP	PVC	BUR	EPDM	TPO	EVA	ECB	PIB	CPE	CSPE
1998	14,479	9,535	4,203	2,759	891	576	430	474	316	105	59
2000	14,516	13,316	4,402	2,349	899	666	414	419	273	89	39
2001	14,213	10,551	4,650	2,132	916	726	433	420	276	89	35
2002[a]	14,271	10,606	4,815	1,965	949	788	454	427	279	89	31
2004[a]	14,669	10,524	5,135	1,705	1,010	926	495	434	282	87	28

Note
a Estimate, abstracted from industry reports.

on the European markets. SBS PMA is the clear leader. Also included are volume data on ECB (ethylene copolymer bitumen) and EVA (ethylene vinyl acetate) that are little known or used in the United States and Canada.

QUESTIONS

1 The US roofing market in 2001 is approximately.... [a] $30 billion [b] $30 million [c] 30 billion squares [d] $25.3 billion.
2 Low-sloped roofs are designed to be watertight and have a slope of 25 percent or less to drains. [a] true [b] false.
3 Steep-sloped roofs are water shedding; they are not watertight under ponded water. [a] true [b] false.
4 Low-sloped roofs represent about... of the roofing sales in 2001. [a] half [b] one third [c] two thirds [d] 60 percent.
5 Cover boards should be used over all foam insulations where the low-sloped roofing is to be adhered. [a] true [b] false.
6 The installations of polyisocyanurate foam, wood, and perlite board insulations for low-sloped roofing systems in 2001 represent about... percent of all installations. [a] 25 [b] 47 [c] 80 [d] 93.
7 EPDM is a steep-sloped roofing product. [a] true [b] false.
8 Architectural metal roofs should be installed on low-sloped decks. [a] true [b] false.
9 Most of the low- and steep-sloped roofing in 2001 was re-roofing. [a] true [b] false.
10 Traditional asphalt built-up roofing represented about... of the new construction low-sloped roofing market. [a] one eighth [b] half [c] 20 percent.
11 EPDM was used for re-roofing about one quarter of the time in 2001. [a] true [b] false.
12 The steep-sloped roofing market is dominated by.... [a] EPDM [b] PVC [c] wood shingles [d] asphalt-glass fiber shingles.

2 Low-sloped roofing systems and materials

GENERAL

Low-sloped roofing systems command the biggest share of the roofing market in the United States and Canada; they also involve the greatest number of systems and a more diverse group of materials than steep-sloped roofing. Both the principal systems and the membrane materials are discussed in the next two chapters. The thermal insulation used with the roofing membrane is an important part of each low-sloped system; the insulation options will be discussed in Chapter 4.

DURABILITY AND CLIMATE

Roofing system durability is the result of many parameters. It is not a single number, but rather a range of performance. Based on a recent roofing industry survey (Cash 1998: 119–24) in a typical hypothetical case, I would expect a roofing system with average or mean life of 20 years (half of the roofs installed in a given area survive for 20 years), to have a minimum life of 11 years (99 percent of the roofs installed in the same area would survive longer than 11 years), giving a durability range of 20±7.4 years. We would therefore expect the life of the system to range from 12.6 to 27.4 years, 95 times out of 100 exposures.

Table 2.1 shows the estimated national average, minimum and 95 percentile range of the service lives of currently popular low-sloped roofing systems. Note that the 20-year life frequently expected by many owners is seldom achieved by an average system; it is only approached at the upper end of the 95 percentile range. While these numbers may be disappointing to many readers, they must be considered to represent remarkably good service considering weather ravages, abuses, and neglect which the typical roofing system is subjected to. These service-life estimates are for properly designed and installed roofs. Improper design or installation can decrease the estimated life by a factor of 2 or 4 times, depending on the defects.

Table 2.1 Average and minimum years of service life, low-sloped roofing systems

System type	Average	Minimum	95%	Range
Asphalt-glass fiber BUR	16.7	9.1	10.3	23.1
SBS polymer modified asphalt	15.9	8.4	9.6	22.2
Poly [vinyl chloride]	13.8	6.5	7.6	20.0
EPDM rubber	14.2	7.0	8.1	20.3
Asphalt-organic felt BUR	14.7	7.3	8.5	20.9
Spray polyurethane foam	12.1	4.8	5.9	18.3
Thermoplastic polyolifin	12.7	6.0	7.1	18.3
APP polymer modified asphalt	13.7	7.1	8.1	19.3

The local climate, since it controls the thermal history of the membrane, has quite an influence, secondary only to the degree of drainage, on the average durability of the system. Figure 2.1 shows a three-dimensional graph of horizontal roof temperatures for each hour of the day and each day of the year. The illustrated temperatures are normal averages for Minneapolis, Minnesota for gray roof; this shape is typical for all locations that we have tested. These data are sinusoidal (smooth waves) in both the hourly axis and the daily axis. This means that the average temperature is a fair representation of the local climate, because the average falls directly in the middle of these data.

We are using the average air temperature as a prime indicator of the local climate. The higher the average air temperature at a location, the shorter the service life of the membrane, because air temperature is correctly considered a result of changes in solar radiation, wind, rain, cloud cover, and all other environmental influences. The average temperature takes into account the sum of all the thermal variations and time. We use the Kelvin

Minneapolis, Minnesota

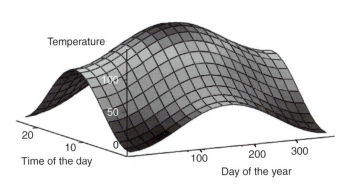

Figure 2.1 Horizontal roofing surface temperature graph.

scale to express the average temperatures because absolute temperatures are required for using the Arrhenius life relationship calculations. The temperature, in °C/°F can be calculated from the temperature in Kelvin (K) by:

$$K - 273 = °C$$
$$(9[K - 273]/5) + 32 = °F$$

Thus,

- 280 K = 7 °C = 44.6 °F, the average temperature of southern Canada and the northern United States,
- 290 K = 17 °C = 62.6 °F, a typical average temperature for the central United States,
- 300 K = 27 °C = 80.6 °F, a typical average temperature for many southern United States, and
- 310 K = 37 °C = 98.6 °F, a very tropical average temperature that is experienced only rarely in the United States in very localized areas.

Table 2.2 shows how the average durability of many low-sloped roofing systems varies with climate. Some systems, such as asphalt-glass fiber built-up roofing, SBS polymer modified asphalt, and asphalt-organic felt built-up roofing show relatively little decline in durability with increasing thermal load, while plastic and rubber systems show significant declines in average durability as the average temperature increases. This indicates that care should be taken in specifying thermally sensitive membranes in warm climates.

Table 2.3 shows the average percentage of surviving membranes after selected exposure intervals, and the estimated life cycle costs for each type of membrane. Note that few, if any, systems can be expected to survive past 27 years. Also apparent, the estimated life cycle costs determined from our survey show these systems are very cost-competitive. Any differences are more likely due to calculation errors than to reality. The highest life cycle cost,

Table 2.2 Mean years of durability at various thermal loadings

Membrane type	Thermal load, K			
	280	290	300	310
Asphalt-glass fiber BUR	18.5	17.4	16.5	15.7
SBS polymer modified asphalt	16.3	15.9	15.5	15.2
Poly [vinyl chloride]	26.8	16.4	10.3	6.7
EPDM rubber	20.1	15.4	12.0	9.5
Asphalt-organic felt BUR	16.9	13.6	12.4	11.4
Spray polyurethane foam	31.8	13.9	6.5	3.2
Thermoplastic polyolifin	14.0	12.8	11.8	10.2
APP polymer modified asphalt	17.3	14.3	1.9	10.1

Table 2.3 Mean percent surviving and life cycle costs

Membrane type	Percent surviving after (in years)					Life cycle cost $/yr
	9	15	21	27	33	
Asphalt-glass fiber BUR	99.9	69.9	9.5	0.1	0.0	0.31
SBS polymer modified asphalt	98.3	61.0	5.7	0.0	0.0	0.34
Poly [vinyl chloride]	93.7	35.2	1.2	0.0	0.0	0.36
EPDM rubber	95.6	39.7	1.4	0.0	0.0	0.33
Asphalt-organic felt BUR	96.4	46.4	2.3	0.0	0.0	0.33
Spray polyurethane foam	83.7	17.9	0.2	0.0	0.0	0.47
Thermoplastic polyolifin	90.0	21.2	0.2	0.0	0.0	0.37
APP polymer modified asphalt	95.3	32.3	0.5	0.0	0.0	0.34

for spray applied polyurethane foam contains a cost for thermal insulation, not present in the other costs.

BUILDING OCCUPANCY

Paraphrasing Roger Bonifont of Ruberoid (UK), the durability of a roofing system relied on the tenderness of its birth, the care in its upbringing, and the frequency of attention during its final years. To be sure, thoughtful selection of a system appropriate for the exposure, knowledgeable and careful installation, and effective maintenance all influence the position of the individual system within the durability range. Our goal then should be to maximize the potential life of every roofing system we design, purchase, or install by appropriate system selection, careful design and coordination of all of the flashing details, assuring the use of the correct order of construction, and providing for the maintenance required.

Many of the factors influencing durability, such as climate, are beyond our control. Other factors, such as the use or physical abuse to which the system will be exposed, are requirements mandated by the specific building in question. Response to these needs is where the thoughtful selection of the roofing system becomes important. For example, it makes little sense to provide a thermally sensitive roofing system in a tropical climate. It is equally inane to provide a soft system, with low static and dynamic impacts, to a building in a climate where severe hail can be expected, or where the roof is used as a workshop, classroom, or where it is subject to extensive foot traffic. It is insane to provide a roof that can be dissolved or otherwise destroyed by the thermal insulation selected, or by the fumes from the manufacturing operations in the building. On the other hand, it makes no economic sense to over specify; to select a roofing system too good for the use intended, although over specification is the least severe error.

Avoid the following in roofing system selection:

- Selecting roofs whose most important attribute is that they are "new" – remember:

When you specify,
You should remember that most
Experiments fail!

- Selecting based on the warranty. I had the daunting experience of having an agent of the Federal government inform me that the warranty was the most important part of the roofing system. What nonsense! No one ever wrote a warranty to protect anyone else. A warranty usually covers about 216×279 mm ($8\frac{1}{2} \times 11$ in.), and is not very waterproof. The period of the warranty has no correlation with the service life of the roofing system. About the only useful function of a warranty may be the legal link it provides to the manufacturer in case of failure, but we would prefer and plan that the roof performs beautifully, and therefore the warranty is redundant and an economic waste.

Warrantees and snow
Have in common that they
Both change to water

- Selection based on the frequently heard phrase: "I've got 20 years experience" by anyone. Remember, the design professional has the task and obligation to select the appropriate roofing system. An effective designer does not blindly do the same thing every year; hopefully, the designer learns from his experience.

Twenty years experience
Some folk have one year – 20 times
Others learn from theirs!

- Many owners and government agencies require designers to provide generic specifications. This gives rise to specifications such as: "Provide a 20-year asphalt, coal-tar pitch, or single ply roof – or equal." This has the effect of abrogating the fundamental responsibility of the professional designer to select the appropriate roofing system. It means that the owner becomes the designer in this area where it is presumed he is less informed and qualified to make the choice. The intent, to provide competition, is laudable, but having experienced and reliable contractors bidding clearly defined specifications better achieve that goal. Ill-defined specifications leave a great deal of room for error that often can cost more than any savings from competitive bidding.

Or equal clauses
Are disaster invitations
Define the job needs!

MEMBRANE PHYSICAL PROPERTIES

Examining the physical properties of roofing membranes can help our selection process when we know the occupancy of the building and the extent of use which the roofing system must withstand. But be careful! The test methods used to test the rubber systems are different from those used to test asphaltic systems, which are different from those used to test PVC and other single ply systems. The manufacturer's test data can provide a comparison within each type of roofing, but they are not valid for the comparison across different types of membranes.

Table 2.4 lists typical fundamental average physical properties for six of the most popular low-sloped roofing membranes. Samples of each of these membranes were subjected to identical tests, performed in a single laboratory, and by a single operator. These data were collected as part of a much larger program to measure how these membranes behave after heat conditioning, after concentrated ultraviolet exposure, and ultimately after six years of outdoor exposure at three locations about the United States. A report, from which these data on Table 2.4 were obtained, was recently published (Bailey 2002).

Let's examine these data, in so far as tensile strength is concerned. Except for the low values obtained with EPDM, most of these membranes are similar to each other. The tensile strength is the product of the reinforcement used, and EPDM is the only membrane in this series of tests that does not include fiber reinforcement. Reinforced EPDM is currently on the market, but was not included in this program.

A tensile strength of 35 kN/m at −18 °C (200 lb/in. width at 0 °F) has been suggested as a lower limit for bituminous membranes (Mathey 1974). The minimum tensile strength for single ply membranes is set by the individual standards for each membrane.

The type of reinforcement influences the magnitude of the elongation illustrated by these samples. The PVC membranes have the best elongation of the reinforced membranes. As could be expected, the elongation shown by the EPDM membranes exceeded the capacity of our environmental chamber, when we needed the chamber to be tested at low and high temperatures, and averaged ~523 percent at room temperature. This does not mean that an EPDM roof is many times superior to the other membranes, because this elongation will be severely reduced by the fasteners used, or by the adhesives and the substrate used, to install the membrane.

As an example: a high strength rubber membrane, with an elongation in excess of 500 percent, was adhered to a plywood deck with a very strong

Table 2.4 Typical physical properties of unexposed low-sloped roofing membranes

Property	Temperature	Units	Membrane					
			Asphalt BUR	SBS-PMA	PVC	EPDM	TPO	APP-PMA
Tensile strength	-18 °C	kN/m	38.2	42.2	36.6	11.4+	37.8	47.8
	23 °C	kN/m	22.8	21.2	27	8.7	11.4	25.8
	70 °C	kN/m	4.8	6.7	13.4	5+	3.2	6.2
Elongation	-18 °C	%	6.2	8.2	34	278+	31	8.3–26
	23 °C	%	4	13–70	36	523.5	23	11–89
	70 °C	%	7.7	14–21	28.5	282+	5.4	7.7–23
Energy to first peak	-18 °C	kN/m	4.4	6–85	24	72+	29	10–30
	23 °C	kN/m	2.2	5–64	19	98	6.8	8–60
	70 °C	kN/m	0.7	7.5	5–11	28+	0.4	1–5
Water absorption	23 °C	%	2.8	1.5	2.6	2.8	6.3	1.8–2.5
Glass transition	–	°C	-21	-44	-44	-52	-36	-32
Thermal expansion		$\times 10^{-5}$/°C	21.3	33.6	9.8	9.8	7.5	29.6
Static impact	-18 °C	N	250	250	250	250	250	250
	23 °C	N	88	98–250	250	250	250	98–250
	70 °C	N	72	35–250	250	250	250	47–250
Dynamic impact	-18 °C	J	6.8	13–21	6.8	5.6	3	15.8
	23 °C	J	5.6	13–18	4.3	3	5.6	13.1
	70 °C	J	4.3	3–8	5.6	3	8.1	4.3

adhesive. In a short time the rubber was split over every joint between plywood panels. The 500 percent elongation means the rubber could extend five times the original length of the unstressed sheet. The original length was zero where the plywood panels were tightly butted and five times zero is still zero. The rubber never had a chance. The only way this assembly could perform would be to gap the plywood (3 mm [⅛ in.] is typical) and perhaps use an adhesive with a lower bonding strength so it could pull free at each joint to increase the original length.

The energy-to-peak strength is considered by many to be important, because it is the work needed to extend the membrane to its limit. The membrane cannot return to its original condition when it is extended beyond this peak. The data in Table 2.3 show that the energy-to-peak is not proportional to either the tensile strength or elongation, but, not surprisingly, can be estimated by half of the product of the tensile strength and the percent elongation, since it is the area under the load–strain curve.

Water absorption is another significant property for roofing membranes. It is particularly useful for built-up roofing membranes, because it can be used as a measure of the relative health of the membrane. Table 2.5 lists the mass and the equilibrium moisture content at 45 percent (EMC_{45}) and at 90 percent (EMC_{90}) relative humidity of most of the built-up roofing materials. To calculate the moisture that one could expect in a healthy membrane, multiply the mass of each component by the relevant EMC, add the products together, and divide by the total membrane mass.

As an example: the built-up membrane in Table 2.3 is composed of three plies of asphalt-glass fiber felt that has a mass of 340 g/m^2 (7 lb/100 ft^2) and is adhered together with 1.5 kg/m^2 (31 lb/100 ft^2) of asphalt per ply. From Table 2.4, the EMC_{45} is 0.9 percent and the EMC_{90} is 1.1 percent for asphalt-glass fiber type VI felts. Therefore, the moisture we would expect in the membrane is:

$$EMC_{45} = \frac{(3 \times 340 \times 0.9)}{[3 \times (1500 + 340)]} = 0.17$$

and

$$EMC_{90} = \frac{(3 \times 340 \times 1.1)}{[3 \times (1500 + 340)]} = 0.20$$

Thus, any asphalt-glass fiber felt membrane that contains between 0.17 and 0.20 percent moisture is damp, and any similar membrane that contains more than 0.20 percent moisture is wet; it can provide moisture to the surrounding materials. Other definitions of "wet" have been proposed, and will be discussed in Chapter 3.

Examining moisture absorption data in Table 2.4, it is evident that all the membranes are probably wet after this test, due to water trapped on the

Table 2.5 Roofing material mass and equilibrium moisture content

Common name	ASTM		Dry mass		EMC mass %	
	Designation	Type	g/m²	lb/100 ft²	45% RH	90% RH
Asphalt-cotton fabric	D173		340	697	3.7	5.5
Smooth roll roofing	D225	I	1943	39.8	1.4	2.7
		II	2666	54.6	2	3.8
		III	2495	51.1	2	3.8
		IV	1943	39.8	1.4	2.7
Asphalt-organic felt	D226	I	560	11.5	4.3	8.2
		II	1270	26	4.1	7.9
		III	830	17	4.3	8.2
Coal tar-organic felt	D227		635	13	4.3	8.2
Asphalt coated and granule surfaced – 90#	D249	I	3610	74	1.5	2.8
		II	3490	71.5	1.5	2.8
Asphalt-asbestos felt	D250	I	630	13	1.7	2.7
		II	1370	26	1.6	2.6
		III	830	17	1.7	2.7
		IV	1030	21	1.8	2.8
Asphalt and granule coated – wide selvage	D371	I	1806	37	1.7	3.2
		II	2260	46.3	2.1	4
		III	1733	35.5	1.7	3.2
		IV	2090	42.8	2.1	4
Asphalt-burlap fabric	D1327		330	677	14.8	23.4
Asphalt-glass fabric	D1668	I	65	1.4	1.4	1.6
Coal tar resin-glass		II	69	1.4	1.3	1.5
Organic resin-glass		III	50	1.1	1.8	2.1
Asphalt-glass felt	D2178	IV	342	7	0.5	0.6
		VI	342	7	0.9	1.1
Asphalt-organic base sheet	D2626		1806	37	1.5	2.9
Asphalt-coated ply sheet	D3158		1420	29	1.9	3.7
Asphalt-asbestos base sheet	D3378	1	1810	37	0.6	0.9
		2	1900	39	0.7	1.2
Asphalt-asbestos vent felt	D3672	1	2930	60	0.4	0.6
Asphalt-glass vent felt		2	2440	50	0.1	0.1
Asphalt-glass #90	D3909		3085	63.2	0.1	0.1

surfaces and water wicked into the membrane. The lower the water content in the membrane, the better. As measured by this test, any water absorption greater than three percent should be suspect in any membrane, because it implies a penetration of water into the membrane – usually along the reinforcing fibers. Severely deteriorated built-up roofing samples can contain four to more than 20 percent water, based on the dry weight of the membrane.

Rubber membranes are not immune from internal water damage. One manufacturer changed from a lead to a magnesium catalyst without informing anyone of the formulation change, and perhaps without investigating the water absorption of the newer product. The new product, unlike the old product, swelled due to water absorption, and opened the new seams before they had a chance to properly cure. Water poured into the building and cash poured out of the manufacturer's coffers to replace the roofing and to pay for the consequential damages.

The glass transition temperature (the point when it changes from a brittle solid to a flexible solid) is obviously of interest. This parameter depends on the polymers involved, rather than the reinforcements used with the membrane. As reported in Table 2.3, the glass transition points of these common membranes range from −21 to −52 °C (−6 to −62 °F). Little is known about the change in glass transition temperature with outdoor exposure; it is suspected that they might rise, making the membranes harder and less flexible. This is currently being investigated.

We measured the thermal expansion coefficient for these membranes, and plan to explore if and how this property changes with outdoor exposure. The asphalt containing membranes have roughly three times the movement of the plastic and rubber membranes. The thermal expansion coefficients for the asphaltic membranes are $210–340 \times 10^{-6}/°C$ ($380–610 \times 10^{-6}/°F$). The thermal expansion values for the plastic and rubber membranes are $80–100 \times 10^{-6}/°C$ ($140–180 \times 10^{-6}/°F$).

Static and dynamic puncture tests measure the ability of the membranes to resist puncture under controlled laboratory conditions, using the same type of substrate for each membrane. Data in Table 2.4 show that the asphaltic membranes are not as resistant as the plastic and rubber membranes to static impact. The polymer modified asphalt membranes (or more correctly, the reinforcements used) are more resistant to dynamic impact than the plastic or rubber membranes.

DRAINAGE

The next critical step, after the selection of the roofing system appropriate for the climate and the occupancy of the building, is to make sure that all roofing and flashing surfaces drain promptly. The fundamental function of a roof is to channel precipitation off the roof and around the building envelope. Roof areas that promptly drain last at least twice as long as areas that don't drain promptly. Current wisdom suggests that the drainage is adequate if the surface does not pond water 24 h after a rain. I would prefer to see no ponds on the roof surface at any time, including during the rain. Remember that you can only stop water permanently with very great difficulty (it made the Grand Canyon); channel the water to the drains and away from the building.

Roofing is simple
Slope all surfaces to drains
Or they will leak.

Effective drainage makes the single greatest contribution to the durability of the roofing system, regardless of the system selected. In northern climates, drain leaders should have a minimum diameter of 100 mm (4 in.) to minimize blockage from ice. Ideally, the slope of the structural deck should provide the slope to drains, because it is usually lower in cost than providing slope with tapered insulation. The drain in each area must be properly sized to accept the drainage. Table 2.6 provides some guidance, but the local building codes usually control the drain sizing.

Divide the roof surface into smaller square or rectangular areas to use more rather than using larger drains. I am not aware of a standard for the size of drainage areas, but prudence suggests a good maximum is about 2000 m^2 (~200 squares). It is good practice to use two drains in each drainage area in case one drain strainer becomes clogged. Some building codes require two separate drainage systems for each roof section to protect against plugged storm piping and the potential overloading of the structure by ponded water. A slightly raised metal edge flashing can serve as secondary drainage, should the primary system become clogged, and is much less expensive than a dual drainage system if it is permitted by the building code

Table 2.6 Roof drainage guide

Rain fall intensity (for 5 min)		Drain capacity[a]		Corrections for sloped roofs		
mm/min.	in./h	m^{2}roof/m^{2}drain	ft^{2}roof/ in.^{2}drain	Slope (%)	Slope (in./ft)	Area factor
0.85	2	86,400	600	0–25	0–3	1.00
1.27	3	57,600	400	33–42	4–5	1.05
1.69	4	43,200	300	50–67	6–8	1.10
2.12	5	34,560	240	75–92	9–11	1.20
2.54	6	28,800	200	100	12	1.30
2.96	7	25,200	175			
3.39	8	21,600	150			
3.81	9	18,720	130			
4.23	10	17,280	120			
4.66	11	15,840	110			

Source: Adapted from industry literature.

Note
a For residential roofs use one square metre of drain for every 14,400 square metres of roof area (one square inch of drain for every roofing square of corrected area).

and the local building inspector. In a similar manner, a scupper can provide emergency drainage, where the roof has parapets. These are both discussed in Chapter 5, on flashing.

Note that this maximum suggested size is about the greatest area that can be roofed by a good crew of roofers in one day. Half that size is the production rate for a typical crew. For this reason, I would prefer to see the drainage area limited to about 1000 m² (~100 squares). By limiting the size of each drainage area, the designer can try and assure that the work of each day has proper drainage and that the differential between the elevation of the drain and the perimeter is minimized.

Consider a simple example of a drainage area of 1200 m², 30×40 m, with a central drain. The minimum slope, selected to be two percent (¼ in./ft), will be located along the 25 m long diagonal of the rectangular area from the corner to the drain. With the drain elevation set at zero, the perimeter elevation is 0.02×25 = 500 mm. This gives a slope to drain of 0.5/15 = 3⅓ percent (0.4 in./ft) in the shortest distance to the drain, and a slope of 0.5/20 = 2½ percent (0.3 in./ft) perpendicular to the shortest slope. Using the structural deck to provide the needed slopes is much lower in cost than building these slopes up to a thickness of 500 mm (~20 in.) of thermal insulation at the perimeter. Obviously, quartering the drainage area to a 300 square metre area results in an elevation differential of 250 mm (~10 in.) at the drain and the perimeter.

ASPHALT LOW-SLOPED ROOFING MEMBRANES

Asphalt membranes such as built-up, polymer modified, and cold process represent over one-third of the current low-sloped roofing market. Asphalt and coal-tar pitch, being the oldest of the thermoplastics used in roofing, deserve some discussion.

Asphalt and coal-tar pitch

Asphalt is a complex blend of hydrocarbons too numerous to mention. It originates as the controlled residue of the distillation of asphaltic petroleum. About 15 percent of the petroleum asphalt is used for roofing. By far the larger share, 85 percent, is used for road building. Asphaltic petroleum is found in many locations. Notably, asphaltic petroleum is found in Alaska on the North Slope (Abraham 1945), in the mid-continent area such as Kansas and Texas, and in Venezuela. California and Pennsylvania crude petroleum are paraffinic. This makes them very suitable for lubricants, but less suitable for roofing asphalt.

Asphalt also exists as a product that can be mined. Major deposits exist in Trinidad (Trinidad lake asphalt was used in Washington, DC to provide the first asphalt pavement) and in the Dead Sea in the Near East. Dead Sea asphalt may be the reason that the Babylonian civilization was able to develop in an area that experienced annual floods. They used a mixture of Dead Sea asphalt and straw as mortar for constructing walls of sun-baked bricks that could resist the floods. Asphalt was one of the materials used by Egyptians to prepare mummies. They also exposed water-filled, asphalt-coated wooden trays to the night sky to make ice. Very hard asphalts are found in Oklahoma and Utah. These natural mined asphalts are used for many special coatings, inks, and other specialty compounds, but few of these are used in roofing, largely because the petroleum asphalts are so inexpensive, can be manufactured to the viscosity required for each application, and with the quality control we currently demand.

The asphalts used in built-up roofing are described in ASTM D312 *Standard Specification for Asphalt Used in Roofing*. There are three grades that differ primarily by softening point and viscosity. Softening point is the temperature of a water bath, measured when a 10 mm (⅜ in.) diameter stainless steel ball slumps 25 mm (1 in.) through a 19 mm (¾ in.) diameter, 6.4 mm (¼ in.) thick ring filled with asphalt, when the water bath temperature is raised at the rate of 5 °C (9 °F)/min. The test method is fully described in ASTM D36 *Standard Test Method for Softening Point of Bitumen (Ring-and-Ball Apparatus)*. Softening point is a crude viscosity test; it has been used quite effectively for many years as a quality control tool.

Determining the viscosity at various temperatures requires more elaborate equipment. The viscosity equipment most frequently used today is described in ASTM D4402 *Standard Test Method for Viscosity Determinations of Unfilled Asphalts Using the Brookfield Thermosel Apparatus*. Newer and much more elaborate viscosity measuring methods have been developed in the paving side of the industry, but have not yet gained acceptance in roofing.

The asphalt viscosity–temperature relationship is linear when the logarithm of the viscosity is graphed with the temperature. This relationship varies with the type of asphalt studied. For asphalt application on the roof, we are interested in two points on this viscosity curve. These are called the equi-viscose temperatures (EVT) for hand mopping and for the machine application of the asphalt. The EVT for hand mopping is the temperature at which the asphalt's apparent viscosity equals 125 centipoises. It is 75 centipoises for machine application. Coal-tar pitch also has an EVT; it is the temperature at which the pitch's apparent viscosity is 25 centipoises. The range for each EVT is ±14 °C (±25 °F). The application rate of the bitumen is expected to be appropriate if the asphalt or coal-tar pitch is applied within their respective range.

We also must be interested in the flash point of the asphalt (it is measured using ASTM D92 *Standard Test Method for Flash and Fire Points by Cleveland Open Cup*) because of our concern about fire. Most roofing asphalts have a flash point in excess of 260 °C (500 °F). This is only slightly higher than the carbon–hydrogen excitation temperature of about 250 °C, where asphalt starts to degrade (482 °F) (where the vibrating carbon and hydrogen atoms tend to fly apart), and quite close to the average temperatures encountered in the kettle where the asphalt is melted.

On one occasion, a kettle man moved his kettle into an unfinished building because a sudden rain storm threatened. The other workers inside complained of the fumes, so the kettle man rigged up an electric fan to blow the fumes out of the building. The asphalt was overheated since it was not being used, and the air forced by the fan over the open kettle was enough to provide spontaneous combustion of the asphalt. The building was quickly involved and finished. Not quite the way the owner planned.

The viscosity of the asphalt at temperatures that are normal on the roof is important, because we must be concerned about the possibility of the roof sliding after application. Roof top temperature viscosities can be estimated using ASTM D4989 *Standard Test Method for the Apparent Viscosity (Flow) of Roofing Bitumens Using the Parallel Plate Plastometer*. Currently, studies are continuing to attempt to grade roofing asphalt by viscosity, and to set minimum viscosities for each grade near normal roofing surface temperatures. Once completed, compliance with the standard will tend to eliminate problems related to the sliding of the components.

Asphalt is sometimes erroneously called "tar," as in: "tar-paper." Technically, "tar" is the distillate resulting from the dry distillation of organic materials. "Pitch" is the residue of the fractional distillation of tar. "Stearin pitch" is the residue of the distillation of tar, obtained by the distillation of animal fat. In a somewhat similar manner conceptually, coal tar is the distillate of the dry distillation of coal that includes naphthalene (moth balls) and other valuable natural chemicals. The residue of the dry coal distillation is coke, much valued in the steel industry as a source of carbon. The residue of the fractional distillation of coal tar is coal-tar pitch, a valuable type of bitumen, also used in built-up roofing.

While both asphalt and coal tar pitch are used in roofing, they are very different, and should be mixed only in systems approved by the bitumen manufacturers. Coal-tar pitch contains free or elemental carbon filler which is insoluble in most solvents; it helps to provide coal-tar pitch with a much higher density than water. Pure asphalt does not contain fillers and has a density only slightly higher than water. These differences can be used to distinguish between these bitumens. Drop a small piece of bitumen in a container of water. It may be coal-tar pitch if it rapidly sinks to the bottom. Finding a black insoluble residue after washing a small bitumen sample with any available petroleum solvent, assures the presence of coal-tar pitch. The quantity of free carbon in the coal-tar pitch is critical to control the flow of the pitch; without it the pitch might flow off the roof.

Asphalt and coal-tar pitch are incompatible with each other. When they are brought into contact, relatively low molecular weight components move from the asphalt to the pitch, and the combination forms a greasy, grainy mass that can no longer exclude water, showing that the colloidal structure of the bitumens is destroyed.

Coal tar built-up roofs formed with coal tar – glass fiber felts or coal tar organic felts are a very small part of the market. Some very conservative owners demand the same roof their grandfathers used. At some other locations, the prices of the materials are competitive with asphalt prices, including the surcharge required by some unions to remove and install pitch roofs, because of the carcinogenic nature of the fumes from the pitch. (Gone are the days when we picked pitch from the street to use as chewing gum!) The odor of hot pitch can generate demands to stop application if the odors can get to personnel in the building being roofed.

Coal-tar pitch roofs must have a very low slope because of the tendency of pitch to flow into drains, and any hole in the roof and to, if you are not very careful, drip into and stain the interior of the building. Often lauded as "self healing," self healing property of pitch is seldom observed outside the laboratory.

A very serious problem surfaced when pitch was used with asphalt-coated glass fiber felts to form built-up roofing membranes. Forced downward by the gravel surfacing, the felts strained the pitch through their pores, greatly increasing the asphalt-pitch contact and hastening incompatible reactions that destroyed the structures of the bitumens, and permitted extensive general water leakage. Unfortunately, this construction was used on many roofs.

Coal-tar pitch roofing
With asphalt-glass felts give us
Clients and profits.

Glass fiber felt displacement has been observed when low viscosity asphalt was used with porous asphalt-glass fiber felts. In these cases, the felts were pressed to the bottom of the membrane by the gravel surfacing, but incompatibility was not observed.

QUESTIONS

1   Low-sloped roofing has the [a] smallest [b] largest share of the roofing market in the United States and Canada.
2   Roofing durability is [a] a single value [b] a range of values [c] the value of the warranty.
3   The average roof lasts [a] for 20 years [b] for a value depending on the system selected [c] for ever.
4   The local climate has a [a] severe [b] no [c] a moderate effect on roofing system durability.
5   The [a] average [b] highest [c] lowest temperature is the best indicator of the severity of the climate.
6   For calculation [a] the Celsius [b] Kelvin [c] Fahrenheit temperature is used.

7 A typical average temperature for the United States is [a] 80.6 °F [b] 280 K [c] 17 °C.

8 Asphalt built-up roofs show a [a] large [b] small [c] moderate decline in durability as the average temperature increases.

9 The selection of the roofing system is [a] highly [b] not [c] seldom influenced by the exposure to which it will be exposed.

10 "New" is the most important reason for choosing a roofing system. [a] true [b] false.

11 The warranty offered should not be considered in membrane selection. [a] true [b] false.

12 It is very important to specify at least three different roofing systems. [a] true [b] false.

13 Competition is best served by an "or equal" clause. [a] true [b] false.

14 The designer should leave the roofing system selection up to the [a] roofing contractor [b] owner [c] general contractor [d] none of these.

15 The stronger the roofing membrane, the longer it will last. [a] true [b] false.

16 The system's elongation is greatly influenced by the reinforcement and the membrane's attachment to the building. [a] true [b] false.

17 The energy-to-peak is proportional to half the elongation times the load at the first peak. [a] true [b] false.

18 The equilibrium moisture content at 90 percent relative humidity is twice as high as the equilibrium moisture content at 45 percent relative humidity. [a] true [b] false.

19 Roofing membranes at room temperature cannot donate water to the surrounding materials if their moisture content is less than the EMC. [a] true [b] false.

20 The drier the membrane, the better its health. [a] true [b] false.

21 Glass transition temperatures below 0 °C are dangerous. [a] true [b] false.

22 Rubber membranes have a larger apparent thermal expansion coefficient than asphalt membranes. [a] true [b] false.

23 The asphalt membranes are more resistant to static and dynamic impact than the rubber or plastic membranes. [a] true [b] false.

24 Effective drainage has the greatest effect on the life of the roofing. [a] true [b] false.

25 Ideally, drainage areas should be about the area that can be installed in a single day. [a] true [b] false.

26 All major roof areas should have at least two ways by which storm water can leave the roofing surface. [a] true [b] false.

27 Smaller drain areas increases the size of the blocking with sloped insulation. [a] true [b] false.

28 The minimum drain leader diameter suggested is [a] 25 mm [b] 50 mm [c] 100 mm [d] 200 mm.

29 Asphalt and tar are the same. [a] true [b] false.

30 Most of the asphalt sold each year is for road building. [a] true [b] false.

31 Asphalt from crude oil is preferred for roofing over mined asphalt because we can more accurately control its viscosity. [a] true [b] false.

32 Softening point is a measure of asphalt viscosity. [a] true [b] false.

33 Asphalt's viscosity is linear when graphed against temperature. [a] true [b] false.

34 EVT for hand mopping is the temperature where the asphalt has an apparent viscosity of 125 centipoises. [a] true [b] false.

35 Asphalt starts to degrade when it is heated over 212 °F. [a] true [b] false.

36 The parallel plate apparatus can measure the apparent flow of an asphalt at roof top temperatures. [a] true [b] false.

37 Coal-tar pitch is denser than asphalt and contains free (insoluble) carbon. [a] true [b] false.

38 Coal-tar pitch should always be used with asphalt-glass fiber felts. [a] true [b] false.

3 Low-sloped roofing systems and materials (continued)

ASPHALT-GLASS FIBER BUILT-UP ROOFING

Asphalt-glass fiber built-up roofing membranes are composed of alternate layers of glass fiber felt and relatively high melting point asphalt. Three to four layers of glass felts are often used. The top surface is coated with asphalt, asphalt emulsions, or asphalt and gravel or slag. The gravel functions as ballast to hold the roof in place, to protect the asphalt from degradation due to radiation, and to improve the system's fire resistance. The asphalt and gravel finish is most frequently used. Properly adhered or fastened gravel surfaced built-up roofs are seldom displaced by the wind, unless the structure that supports them fails. Select an aggregate that complies with ASTM D1863 *Standard Specification for Mineral Aggregate Used on Built-Up Roofs*. Try not to use relatively fine gravel such as pea gravel. Pea gravel is too easily displaced by wind scour, and if applied at ~20 kg/m² (~400 lb/100 ft²), much more than 50 percent of the surfacing will be loose and subject to wind scour. The aggregate's value as ballast is decreased if the quantity of pea gravel applied is reduced.

Gravel surfaced asphalt-glass fiber felt built-up roofs are suitable for use almost anywhere a low-sloped roof is required. They are not suitable for use if the roof is going to be exposed to organic acids, such as food processing facilities that dump waste on the roof, or for small and hard-to-reach roofs. Use coal-tar pitch, PVC, or other membranes that are chemically resistant to the food processing debris, although a better plan might be to prevent the debris reaching the roof – whatever its composition. Small or hard-to-reach roofs are economically handled when the materials can be easily carried up a ladder, or transported in an elevator. Roof decks with a slope in excess of 12 percent (3 in./ft) should not use hot asphalt built-up roofing because of the difficulty in controlling the asphalt at these slopes, and the increased danger to the roof applicators.

Given suitable support, a properly constructed gravel surfaced asphalt built-up roof is quite resistant to hail and similar occasional impacts. It requires protection with pavers or elevated platforms if it is going to be constantly used as a working surface.

Built-up roofing membranes are usually fully adhered to the substrate with hot bitumen. Do not attempt to adhere built-up roofing directly to styrene foam insulation because at its proper temperature, the hot asphalt will melt the foam. Use a cover board, by mop-and-flop application, to allow the asphalt to cool slightly before it reaches the foam. This application technique requires relatively small pieces of cover board such as 610×1220 mm (2×4 ft), because larger sheets coated with hot asphalt are too dangerous to flip over. Cover boards are also required over polyisocyanurate foam insulation because of the poor strength between the facer and the foam. See Chapter 4 for a more detailed discussion on polyisocyanurate.

Membrane back nailing is required with steeper slopes. If nails are required, they must go directly into wood elements. The nail-holding power of insulation is too low to permit proper anchorage. Alternatively, the fastener provides too long a lever arm for lateral stability when it is installed through the insulation. If relatively steep roofing is required, run the felts with the slope and provide wood blocking nailers no more than 3.6 m (12 ft) apart to nail off the felts.

Built-up roofs require the following precautions:

- Use glass felts that have tightly meshed fibers and use only steep (ASTM Type III) asphalts. Using low viscosity asphalts and open glass fiber felts induces the felts to migrate within the membrane, to float to the top of membranes without surfacing or to be pressed to the bottom of surfaced membranes. Prudence suggests testing asphalt sampled at the job for softening point, to be sure that the correct grade of asphalt will be applied. Place a sheet of felt over an open newspaper. If you can read the paper through the felt, it is too porous.
- Install only dry materials under dry conditions. Water trapped within the membrane can cause blistering and leakage. Use a thin glaze coating of field applied asphalt whenever rooftop production is interrupted by the weather or poor planning.
- Apply the asphalt at the EVT temperature.
- Seal all edges of the work daily, and remove the seals as work continues, because the sealing materials, such as glazed felts, may have become damp from dew or other precipitation.

POLYMER MODIFIED BITUMENS

Several different types of polymer modified systems have evolved in recent years. The polymers used to modify asphalts are quite diverse. They are all hydrocarbons of substantial molecular weight. They are manufactured by mixing soft asphalt with the polymer selected, at elevated temperatures, using a high shear mixture to disperse the polymer. At the start of the mixing process, the polymer is dispersed in the asphalt. During the mixing, an inversion takes place, where the asphalt becomes dispersed in the polymer. If the asphalt and polymer are correctly matched, the inverted dispersion is quite stable, with the result having a much higher softening point, greater toughness, and enhanced elasticity compared to air blown asphalt of the same stock. If the asphalt is not carefully selected, the modified asphalt may not be stable and may collapse into a soft asphalt and a gooey resin.

Atactic polypropylene (APP) modified asphalt originated in Italy, in the warmer part of Europe. We are generally more familiar with the isotatic form (IPP), or more crystalline form of polypropylene that forms plastic hinges and many automobile parts. APP was a by-product of the isotatic production. At one time it was unusable and unwanted. It was hauled to dumps and buried. More recently, since its use to modify asphalt became commercially important. APP has been the primary product in some plants and IPP the by-product. APP is amorphous (like a gel or liquid). A small percentage of IPP is used in APP modified asphalt to increase its strength and stiffness.

Unlike built-up roofing membranes, with the three or four layers of reinforcing felt, modified bitumen membranes are often formed by two modified asphalt-coated felts; thereby saving some labor costs, theoretically. The coated felts can contain unwoven glass fiber, woven glass fiber, or polyester fiber reinforcing layers. A surface coating can be applied or a roofing granule surfaced sheet can be used, but an APP membrane can be exposed without surfacing. APP-coated felts are most often torched together with propane torches because it is difficult to adhere the sheets with regular built-up roofing asphalt; the APP modified asphalt softening point is too high to properly fuse the sheets together. Torching requires great care even in a professional's hands. It can be a disaster when an inexperienced personnel attempt it. Require a fire watch for at least an hour at the conclusion of each installation break to catch small fires before they become conflagrations. Remember:

> *Torch applied roofing*
> *Is bonnie for APP*
> *But watch out for fire!*

Some roofing contractors are manufacturing SBS (styrene-butadiene-styrene) or SEBS (styrene-ethylene-butadiene-styrene block copolymer) polymer modified built-up roofs by applying polymer modified asphalt (sometimes melted on the roof in the equipment used also to melt rubberized asphalt) with glass fiber mats, polyester felts, or glass fabrics to form the membrane. The SEBS polymer modified asphalt has a greater thermal resistance than the SBS modified asphalt, and can be hand mopped, but any modified asphalt should be heated with care to avoid degradation. These systems are performing very well, but have a limited market due to price, and a history that thus far has not demonstrated the improved performance needed to justify the increased expense.

Styrene-isoprene-styrene (SIS) modified asphalt is softer than the SBS modified asphalt; it is used in pressure sensitive applications. Remember, adhesion in any pressure sensitive application develops from a relatively high pressure for a short time. Unless the adhesive cures or hardens, small pressures for a long time will cause loss of adhesion. As an example: a well-known roofing manufacturer developed a very interesting roofing system that used pressure sensitive tapes over the joints of the roof sheets. The assembly was quite attractive – until thermal cycling caused the tapes to come loose and fall off.

> *Unless glue hardens*
> *Don't use pressure sensitive*
> *Roofing materials.*

Styrene-butadiene-styrene modified asphalt-coated felts can be torched in place, but it is more frequent that these heavy sheets are either mopped in place with ASTM Type VI (super-steep) built-up roofing asphalt or adhered together in a two-ply system with a special cold process asphalt adhesive. The top sheet is usually surfaced with roofing granules similar to those found on asphalt roofing shingles.

The SBS membranes are strong, impact resistant and much more elastic than any conventional built-up roofing membrane. There have been some problems with blistered cap plies (as the top granule-surfaced coated felts are called). Installing the cap sheets in a cold applied adhesive has minimized this problem.

Hybrid roofs are roofs that use two- or three-ply asphalt – glass fiber felt conventional built-up roofing techniques with an SBS polymer-modified asphalt granule surfaced cap sheet. They have been used at many locations recently.

Another recently proposed system is polymer modified coal tar. This is a much newer system that is tarnished by the failure of a previous attempt at modifying pitch. Either of these systems appears to have some merit, but again, their track record is not yet clear as to the benefits obtained, for the risk taken in specifying a new system. Of these two, the hybrid concept is the most conservative; its use represents the least risk.

SINGLE-PLY LOW-SLOPE ROOFING SYSTEMS

EPDM (ethylene-propylene-diene terpolymer)

This is the most popular single-ply roofing system on the market in the United States. The "M" at the end of "EPDM" refers to a methyl linkage, and not another polymer. Currently the major US producers are the "Carlisle Rubber Company" and "Bridgestone-Firestone." Unlike the other single-ply roofing candidates that are thermoplastic, EPDM is thermosetting. It will burn without melting when exposed to fire.

Rubber polymers, carbon black and oils, for processing ease, are compounded, extruded into thin sheets, two sheets are laminated, coated with an anti-stick agent, and the compounds are cross-linked by curing to form a true rubber sheet. The phenol–formaldehyde liquid adhesive used to join the rubber splices in the field, early in EPDM's history, has been supplanted by uncured butyl rubber tape. Properly formed, the rubber tape splices have a much higher early strength – 875–1225 N/m (5–7 lb/in. width) tee peel – than adhesive splices – 175–350 N/m (1–2 lb/in.) tee peel, and also significantly stronger cured splices.

EPDM membranes swell in petroleum solvents, cutting oils, lubricating oils and food greases, and they have low dynamic impact resistance. EPDM's usually very large sheets makes it suitable for use over warehouses, offices and similar roofs that have few penetrations. White EPDM is offered at times. Avoid it; it does not have the carbon black so necessary to protect the rubber from solar radiation and oxidation. EPDM is not indicated in roofs with even moderate traffic or roofs on multi-residential units or schools where sunbathing, barbecuing, partying, sunset watching, or similar regular activities are carried on. Although I approve of all of these activities, I have seen roofs fail because of each of these exposures – the roofing must be protected if the owner is going to permit these activities.

PVC (poly vinyl chloride)

PVC membranes are all currently reinforced. When first introduced to the United States in the early 1970s, the PVC was not reinforced and contained a relatively simple DAP (di-akyl-phthlate) plasticizer. Pure PVC is very hard, stiff and strong. It must be plasticized to be suitable for roofing. DAP is frequently exuded from the vinyl covers on promotional three-ring binders. The problem with the early PVC is that the plasticizer used was fugitive – it left the membrane, upon exposure, age and water soak, just as it leaves the binders on the shelf.

Plasticizer loss causes the plastic to shrink, loose volume, become brittle and finally shatter. In one case, I traced the large number cracks in a ~1300 m² (~14,000 ft²) shattered roof, by tracing the cracks, that extended like branches on a tree, back to the limb, branch and trunk, to the single source of the crack – a dislocation of the perimeter edge metal. Fortunately, most of these roofs have been replaced.

The newer PVC formulations are reinforced with polyester and/or glass fibers and use more complex phthalates and copolymers as plasticizers; these are more resistant to the weather than the straight chain phthalates used earlier. PVC membranes are generally narrower than those of EPDM, permitting them to be used on roofs with many penetrations, and roofs with difficult access. In addition, the PVC seams are heat-sealed or solvent-welded; the seams can therefore be inspected shortly after they are formed – unlike the tape adhered EPDM seams that must wait for the seams to cure before inspection. PVCs are valued for their resistance to most chemicals, and their white color which helps reflect heat off the roofing surface. Be careful when walking on wet PVC roofs; a dew covered PVC is slippery like an ice skating rink; be particularly careful in the shadows of chimneys, air conditioning units, and parapets – where the dew can linger and cause you to slip and fall.

TPO (thermoplastic polyolefin)

It is one of the recent entries to the market. The polymer portion is a blend of polyethylene and polypropylene; the membrane includes pigments, fillers, ultraviolet light stabilizers, antioxidants, and is usually reinforced with polyester and/or glass fabrics. A number of manufacturers have introduced TPO membranes. At least one has already withdrawn their product and another producer (not a major roofing manufacturer) has withdrawn their product and is the subject of a class action suit because of product failures in service. Embrittlement, loss of antioxidants and ultraviolet stabilizers and generalized cracking are reported for this latter system.

The large number of "me too" manufacturers suggest that we are going into a "new product syndrome" with TPO. Whenever a new product or application appears, the following events usually take place:

- Competing manufacturers (those who actually manufacture the product) and private label purveyors (those who purchase the product for resale under their own brand or label) quickly come out with a similar product. They assume the testing of the new material or the originator has successfully completed the system, and they need not bother with any expensive or time-consuming research.

> *To assume usually*
> *Done by one who comes before*
> *The u and the me.*

- In the haste to market, any considered judgment or experience is replaced by wishful thinking – if not downright lies. Some examples: Our product can be installed in any weather – even over snow! Our product is self flashing – no other materials are required! Our product is fool proof!
- Amazing warrantees are offered, with a suggested product or system service life many times the actual experience with the product.

- Roofing contractors are pressured into accepting, recommending and installing the new product or system. With established and experienced contractors, these offerings are listened to – and ignored. They are too conservative, have seen too many of these new systems fail, and don't need to take the risk or the hassle associated with the learning and training associated with a new product. The problem arises when a new contracting firm tries to buy their way into the market by offering roofing at a price below the cost of the other systems available.
- Gullible owners read and believe the literature offered, and decide they know more about roofing than either the roofing design professionals or successful contractors.

Current knowledge suggests using TPOs, if you must, manufactured by major manufacturers, on small roofs. Consider using an inverted roof due to the ultraviolet sensitivity of the membrane. Use only reinforced membranes.

Like PVCs, TPOs offer many potential benefits, including heat-sealed seams that can be checked during assembly, usually light color (although they might benefit from the addition of carbon black), and the potential for low cost.

Other single-ply and low-sloped roofing systems

These include CSPE (chlorosulfanated polyethylene), CPE (chlorinated polyethylene), PIB (butyl rubber), neoprene (chlorinated rubber), sprayed in place urethane foam, and the many liquid coatings used on foam roofing or other substrates. These liquid coatings include: acrylics, aluminized asphalt, asphalt emulsions, silicone rubber, and urethanes.

CSPE systems were originally a liquid-applied top coating for the liquid-applied neoprene/Hypalon® system that may still be found, if infrequently. Hypalon polymer is chlorosulfanated polyethylene; the name is a registered trademark by Dupont. In more recent years, several attempts were made to exploit the excellent weathering properties of Hypalon, including a neoprene-treated asbestos sheet coated with Hypalon, and a glass fiber sheet coated with a blend of uncured Hypalon or Hypalon and PVC, so that the sheets could be solvent or heat welded. Some production and application difficulties of these products have severely curtailed their market.

CPE-coated membranes enjoyed a brief popularity in a small part of the country. The sheet manufactured was thin, approximately 1 mm (0.039 in.), and had a very narrow heat application range. Membrane side lap seals would fall apart if quite enough heat was not provided for heat sealing; holes would be melted in the seam with just a little more heat.

At least one PIB system may remain in the market. The butyl rubber sheet is adhered to a thick polyester fleece, or a mat. The sheets are usually spot adhered to the insulation or the deck. Membrane shrinkage has caused the laps to open on many of these roofs, so the market for PIB roofing has almost disappeared.

Sheet neoprene roofing has been replaced by its lower cost-relative EPDM. There are still sheet neoprene roofs giving satisfactory service at many locations, but they had some cracking problems with ozone exposure, particularly, when the rubber was exposed in a stressed condition.

Polyurethane foam (PUF) roofing, foamed in place has been used for over thirty years on various roofs. Usually these roofs have unusual shapes such as multiple hyperbolic paraboloids, spherical domes, barrel vaults, etc. Very so often we see PUF used to roof over other worn out roofs, and in even fewer cases, as new roofing over poured-in-place concrete and some metal decks.

The foam is generated by mixing an "A" (a polyol) and a "B" part (a catalyst) in a gun and spraying it on the surface where the foam expands and sets to form both thermal insulation and roofing. The foam thickness is usually built-up in 13–25 mm (½–1 in.) thick "lifts" until the specified thickness is reached. Uncoated foam rusts to a brown powder upon exposure, so liquid applied surface coatings are often applied, sometimes with roofing granules (such as the surfacing seen on asphalt shingles) mixed or broadcast into the top coating. Silicone coatings do very well over the foam. Aluminized asphalt or acrylic coatings have a shorter life span and a lower cost.

PUF is sensitive to water (sweat from an applicator's brow will cause blisters) and is the most sensitive to variability in the skill of the applicator. On one occasion, a friend on a new foam roof responded to the question: "How does it look?" with: "Like a herd of dead elephants." The foaming process magnifies any application variance many times. Because of these problems, PUF should only be used as a specialty roofing, and only when a very experienced and skilled applicator is available. Foaming over an existing roof is not indicated because the water inevitably contained in the old roof will cause problems with the foam.

Low-sloped metal roofing

Low-sloped metal roofing is gaining in popularity. This may mean increased future income for building pathologists. These corrugated aluminum or painted steel roofs are often found on "manufactured buildings" and differ from architectural metal roofs in slope and joining details. Side laps are often simply crimped and fasteners in the end laps between metal sheets, at the bottom of the corrugations, frequently ovate out or enlarge the holes in the sheets, by thermal expansion and contraction of the metal, to permit water to leak into the building. Some of the more recent buildings have metal panels that extend from the ridge to the eave of the building, avoiding end laps entirely, but they increase the thermal movement absorbed by the expansion joints in the system. Effective penetration flashing is very difficult in a metal system; both the number and size of any penetrations must be minimized. For example, when two or more penetrations are placed in the same panel, two or more additional points of fixity are provided to concentrate thermal stresses, and to provide paths for water leakage into the interior of the building. Low-slope metal roofing may be a satisfactory answer for temporary buildings with a minimum number of penetrations, but should only be used with caution on buildings expected to give a long service, or temporary buildings that might become candidates for longer service.

ATTACHMENT METHODS FOR SINGLE-PLY ROOFING

Single-ply roofing can be fully adhered, ballasted with rock or pavers or mechanically attached.

For adhered membranes, do not glue them directly to any foam type of insulation because of the mechanical weakness of the foam cell walls. Use a cover board to provide a proper surface for adhesion and to help strengthen the exposed surface. Do not attempt to adhere single-ply membranes (or any other membrane exposed to the weather) directly to concrete decks, because it will bridge over deck imperfections that will be the source of blisters under the membrane. Avoid blistering by using thick polyester felt or a venting asphalt base sheet (for bituminous systems) under the roofing membrane to distribute the vapor pressure of the moisture from the deck.

For ballasted systems, be sure the structure provides the dead load capacity required for the mass of the ballast. The ballast may be a course, round aggregate, pavers, or plastic or rubber walkway pads. Provide a protection board and a geotextile fabric for aggregate ballasted roofs, to prevent sand-wiching the aggregate between the protection board and the membrane. Extruded polystyrene foam insulation can be used as the protection board (many call this assembly, with the insulation over the membrane, an "inverted roof" – I consider it a waterproofing application. Be careful, some manufacturers will not warrant their system in a waterproofing configuration, but may warrant it in an inverted roof). Where concrete pavers are used as a ballast, be sure the pavers are appropriate for the freeze-thaw cycles in the area to which they will be exposed, and be sure they are installed on pedestals or on a drainage board to keep them out of the water. Concrete pavers, even good ones, will disintegrate if they are installed directly on the membrane or the insulation. Some of the newer plastic or rubber walkway pads interlock are massive enough for ballast, and have built-in drainage channels. The waterproofing or roofing system must be drained at the membrane elevation.

Mechanical attachment is not appropriate on concrete decks because it is impractical and can compromise the structural integrity of precast concrete decks – if drilling the holes for the fasteners cut the concrete reinforcing steel. Glass fiber reinforced membranes are required for mechanical attachment to prevent fluttering in the wind.

I have seen standing waves ~300 mm (~1 ft) high, in an EPDM membrane without reinforcement, in a moderate ~19 km/hr (~12 miles/hr) wind. In this case, the fasteners installed within the seam between membrane panels pulled "D" shaped tears in the edge of the membrane, to increase the height of the wind flutter, and so eventual membrane failure.

Whenever mechanical fasteners are in contact with the membrane, some problems can be expected. "The Sun pulls nails" is still sometimes heard. The observation of fastener heads coming through the membrane tends to support that statement, but I can assure you that the Sun does not have sufficient magnetic power to lift a fastener, much less to pull a fastener out of a deck or nailer (else we might have nails floating around in the air, levitated by the Sun). The cause of the "moving fastener" is either the consolidation of the material supporting or around the fastener, such as drying and shrinking wood blocking or ageing, mechanical or destructive consolidation of insulation. Local traffic over the roof with "popped" fasteners results in holes in the membrane and leakage into the building. For this reason, when insulation is mechanically fastened to the deck, a cover board of dense insulation is adhered over the lower insulation layer before the roofing membrane is adhered.

QUESTIONS

1   Gravel is used on a built-up roof to [a] shield against sunlight [b] provide weight [c] increase fire resistance [d] all of the foregoing.
2   Asphalt built-up roofing is very resistant to organic acids. [a] true [b] false.
3   Use cover boards over foam insulation for all adhered roofs. [a] true [b] false.
4   Use low slope asphalt with asphalt-glass fiber felts. [a] true [b] false.
5   Apply the asphalt at the EVT temperature. [a] true [b] false.
6   Leave the nightly seals in place as you continue the roofing application. [a] true [b] false.
7   Polymer modified asphalts are [a] polymers blended in asphalt [b] asphalt suspended in polymers [c] neither of the forgoing.
8   APP modified asphalt-coated sheets are usually torched in place. [a] true [b] false.
9   Pressure sensitive sheets can come apart with a small quantity of pressure exerted in cycles or during a long time [a] true [b] false [c] true, if the adhesive does not cure.
10   The top, cap sheet, or exposed SBS polymer modified asphalt-coated sheet usually is surfaced with roofing granules. [a] true [b] false.

11 Several built-up plies of asphalt and asphalt-glass fiber felt, capped with a granule surfaced SBS polymer modified asphalt-coated cap sheet is called a [a] twin [b] united [c] hybrid roof.

12 EPDM is a true rubber sheet that is thermosetting. [a] true [b] false.

13 EPDM gives excellent service when it is exposed to oils and greases. [a] true [b] false.

14 EPDM is very strong; it resists cutting and abrasion. [a] true [b] false.

15 Early unreinforced PVC membranes shattered after shrinking and becoming brittle. [a] true [b] false.

16 PVC heat sealed seams permit the seams to be quality control checked during installation. [a] true [b] false.

17 Walkers have good traction on wet PVC membranes. [a] true [b] false.

18 TPO membranes have a long history of excellent performance. [a] true [b] false.

19 The warranty period reveals the average service life. [a] true [b] false.

20 PUF roofing is easy to apply; anyone can do it. [a] true [b] false.

21 Low-sloped metal roofing on manufactured buildings must be considered as temporary until such time as the design problems with the thermal expansion and contraction of the metal are overcome. [a] true [b] false.

22 Single-ply roofing can be ballasted, fully adhered, or mechanically fastened. [a] true [b] false.

23 CSPE sheet roofing has been giving trouble free service. [a] true [b] false.

24 PIB roofing is widely used in the United States. [a] true [b] false.

25 Some CPE roofing sheets have a very narrow application range. [a] true [b] false.

26 Fully adhere all membranes directly to foam insulations. [a] true [b] false.

27 The structure must have the added dead load capacity required for ballasted applications. [a] true [b] false.

28 Always use mechanical attachment for the roofing to concrete decks. [a] true [b] false.

29 In inverted or buried membrane roofs, be sure to install drains at the membrane level. [a] true [b] false.

30 Use reinforced membranes for mechanical attachment. [a] true [b] false.

31 Fastener popping is caused by the Sun's magnetic attraction. [a] true [b] false.

32 Traffic over a system with popped fasteners pushes the fasteners back into position. [a] true [b] false.

4   Structural decks and thermal insulation

STRUCTURAL DECKS

The introduction of Factory Mutual Research Corporation (FMRC) is necessary before any discussion of structural roof decks, because FMRC sets minimum standards adopted by the roofing and insurance industries in the United States for steel roof decks and fastening methods for roofing systems. FMRC is an affiliate of Factory Mutual Global; it is located in Norwood, Massachusetts. Their website address is www.fmglobal.com.

Factory Mutual Research Corporation is primarily concerned with the fire, wind, and hail resistance of assemblies to minimize insurance company's risks. Using approved assemblies, including sprinkler systems in some cases, may result in a "Class 1" building, with properly attached roof insulation (for wind resistance), and a roofing system that is hail and fire resistant. "Class 2" constructions cover assemblies that are not rated as "Class 1." "Class 1" constructions are usually lower in insurance cost than "Class 2" constructions.

> *All insulation*
> *Requires firm adhesion*
> *Or it will go bye.*

In re-roofing, one must be careful not to convert an existing "Class 1" construction to a "Class 2" construction inadvertently and add a hidden cost to the re-roofing work by increasing the owner's insurance cost.

Factory Mutual recommendations emphasize steel decks because they are so popular in commercial buildings in the United States and Canada.

> *For steel deck advice*
> *Factory Mutual is nice*
> *Use their service well.*

Important publications include their annual "*Factory Mutual Research Approval Guide*" which includes approval standards information for building materials, and "*Property Loss Prevention Data Sheets*" such as:

- 1–60 Asphalt-coated metal and protected metal buildings.
- 1–28 Wind loads to roofing systems and roof deck securement.
- 1–29 Above deck components.
- 1–49 Perimeter flashing.

A structural deck is the basis for each roofing system. Some decks provide an insulating as well as a structural function (marked with an "I" in the following list) and all decks are classified as nailable or not-nailable (the nailable decks are marked with a "N" in the following list). Some of the decks found in the field include:

- Corrugated steel
- Poured-in-place concrete
- Lightweight structural concrete – I
- Foamed concrete – I
- Vermiculite concrete – I-N
- Poured-in place gypsum – I-N
- Metal-banded gypsum plank – I-N
- Precast, and precast – prestressed concrete
- Concrete – excelsior plank (Insulrock®) – I-N
- Lignin – excelsior plank (Tectum®) – I-N
- Plywood – oriented strand board – N
- Wood planking – I-N
- Wood fiber plank (Homasote®) – I-N.

Of these, *steel decks* are the most frequently used in commercial buildings. These decks are corrugated to form depressed ribs and elevated flanges on 152 mm (6 in.) cycles, 38 mm (1½ in.) deep, or 203 mm (8 in.) cycles, 76 mm (3 in.) deep. The decks are classified by the width of the ribs at the top surface of the deck into narrow rib, intermediate rib, and wide rib configurations. The actual width of the deck rib in each configuration is:

- Narrow rib: 25 mm (1 in.),
- Intermediate rib: 44 mm (1¾ in.), and
- Wide rib: 64 mm (2½ in.).

The decks are also classified as painted or galvanized, and by the thickness of the steel used. The 18 gage [=1.143 mm (0.045 in.)], 20 gage [=0.864 mm (0.034 in.)] and 22 gage [=0.711 mm (0.028 in.)] are most frequently used. The decking panels are spot-welded or mechanically fastened with screws to steel joists or beams. The best practice is to screw the deck to both the top angles of each bar joist and to fasten every sidelap at mid-span, the latter to prevent differential deck deflection response to asymmetric loads.

Table 4.1 lists the maximum spans recommended by the Factory Mutual Research Center for each rib type and deck gage. These maximum spans

Table 4.1 Maximum steel deck panel spans recommended by Factory Mutual Research Co.

Deck type	Steel deck gage		
	18 gage 1.204 mm (0.0474 in.)	*20 gage* 0.91 mm (0.0359 in.)	*22 gage* 0.249 mm (0.0295 in.)
Narrow rib	1.8 m (6 ft 10 in.)	1.6 m (5 ft 3 in.)	1.5 m (4 ft 10 in.)
Intermediate rib	2 m (6 ft 3 in.)	1.6 m (5 ft 5 in.)	1.5 m (4 ft 11 in.)
Wide rib	2.3 m (7 ft 5 in.)	2.0 m (6 ft 6 in.)	1.8 m (6 ft 10 in.)

are conservative enough for most usage. They are 136 kg (300 lb) live load deflection limited to 1/240 of the span, for spans across three supports. Additional supports may be needed for greater loads or fewer supports. Objections frequently heard about sloping the metal decks to drain include: "it is more expensive" or "it requires special bar joist to column connections." These objections are usually offered by designers to the ignorant or lazy to depart from horizontal lines. Special connections are seldom required for bar joists for steel decks sloped up to 4 percent (½ in./ft).

Poured-in-place concrete decks should have their upper surface sloped to drains with adequate allowance in the slope to overcome the long-term creep of the assembly. They are usually used when the interior of the building requires a relatively high degree of fire resistance, such as hospitals, theaters, and high-rise buildings. The top surface should be a light steel trowel finish to enhance adhesion. Rot-proofed treated wood nailers for perimeter and penetration flashing should be bolted to the deck with anchors that are drilled into place. Explosive fasteners should not be used because they can fracture or chip the concrete deck – particularly when their use is attempted near the edge of the concrete. A trap rock aggregate concrete may have a density of 2500 kg/m^3 (156 lb/ft^3).

All concrete decks must be reinforced to minimize cracking. This may seem obvious, but every so often someone decides to install a concrete pad on a roof without reinforcing, for use as a walkway, tennis court, equipment support, pent house, or window washing track over an existing roofing system. The subsequent cracking of the unreinforced concrete tears the roofing membrane apart, permits leakage, and is very difficult, if not impossible, to repair, without removing the concrete.

Lightweight structural concrete often uses kilned shale as an aggregate. It uses less water than foamed or vermiculite concrete, and has a density of 1600 kg/m^3 (100 lb/ft^3). Unless the savings due to lower mass are important, local economics usually makes normal structural concrete less expensive per cubic metre than lightweight concretes.

Foamed concrete is often formed with hydrogen generated by aluminum powder and gypsum added to the sand–cement mix; it uses less water than vermiculite concrete. Foamed concrete can have a mass of about 432 kg/m^3 (27 lb/ft^3) and has a much lower compressive strength than structural lightweight concrete. It has a thermal resistivity "R" of (1.6 in. · hr · ft^2 · °F/Btu).

> *Lightweight is fancy*
> *Stone concrete uses less water*
> *And is less costly.*

Lightweight insulating concrete often uses vermiculite or perlite as an aggregate. Vermiculite and perlite are natural ores that expand upon heating, to form low density aggregate with a very high surface area and ability to hold large volumes of water. The surface of vermiculite or perlite concrete is easily damaged with a carpenter's hammer; it has a density of about 400–432 kg/m^3 (25–27 lb/ft^3). The insulating concrete thermal resistivity is (1.3–1.5 in. · hr · ft^2 · °F/Btu). Insulating concrete is frequently used over dead level steel or concrete decks to slope the surface to drains. Sloping the roofing to drains is wonderful, but the water trapped below the membrane often leads to the roofing system's demise. Lightweight insulating concrete must be vented downward into the building – small metal clips often installed in steel deck sidelaps are not enough venting. Stack vents over the insulating concrete are equally ineffective because there must be a flow of dry air throughout the concrete to dry it effectively.

> *Roof decks that contain*
> *A vermiculite concrete*
> *Must be down vented!*

Poured-in-place gypsum is one of the earliest fire resistant structural decks. Steel bulb-tee (railroad track) shaped sub-purlins are welded across the tops of joists or beams. Wood fiber form board – ~25 mm (1 in.) thick – is installed over the bottom flanges to span the area between the sub-purlins. Steel reinforcing mesh is applied over the sub-purlins, with the mesh sidelaps tied with wire. Hoses are placed perpendicular to the sub-purlins to act as a mold. Water is added to a mixture of plaster-of-Paris and shredded wood fiber, and pumped up and into the volume between the hoses to the desired depth. The hydrated gypsum quickly sets and can be roofed within an hour. The traditional roofing procedure is to nail a base sheet to the gypsum with old fashioned cut nails and metal discs and promptly install a built-up membrane. The cut nails rust in place to improve their holding power.

The structural deck here, like its lightweight insulating concrete cousin, is fire resistant and serves the dual role of insulation and structural support. Gypsum is very fluid during application and is very difficult to slope the upper surface for drainage.

Metal-banded gypsum planks with tongue-and-groove joints are sometimes used for folded plate or monitor roof constructions. The prefabricated galvanized steel-banded planks are installed by welding the bands to clips or directly to the joists. The spaces between the planks at the end joints are grouted. This type of deck must have a layer of insulation to attenuate the deck movement responding to temperature or humidity changes, or these movements can rupture any system directly attached.

Precast and pre-stress concrete panels must have a top pour of reinforced concrete to handle differential movement; an insulation layer alone may not provide enough attenuation. The precast elements are available in a host of shapes such as "single T," "double T," solid or cored plank, and inverted channel. Be sure both the camber and the eventual creep of the deck are taken into account for the drainage plan. Sometimes existing roofs using concrete plank decks, telegraph the floor plan of the building by the separate ponds on the roof over each classroom and corridor. Beware, in this type of construction, of end rotation of the planks; they are sure to rupture the roofing membrane. Also, planks spanning different wall spacing, or planks of different thicknesses, can be expected to respond differently to uniform loading, creating dislocations and destruction where they intersect.

Concrete-excelsior-planks were once marketed under the trade name, "Insulrock," and can still be found in motels and similar constructions. These panels resist combustion and have been used for walls as well as roof decks. The panels are very heavy and are water resistant, but not waterproof. These panels often have tongue-and-groove edges. A typical installation has the ends of panels supported on bulb-tee "rail road rail" sub-purlins and grouted in place.

Lignin-excelsior planks are similar in form to the concrete-excelsior planks, but are much lighter in weight due to the lignin – a cellulose resin – instead of Portland cement used as a binder. The resulting panel responds dramatically to changes in humidity and temperature. Panels under roof leaks tend to sag off their supports into the building. "Tectum" is the trade name for a panel of this construction.

> *Lignin-excelsior planks*
> *Are called shredded wheat planks.*
> *They soften when wet.*

Plywood or oriented strand board prefabricated stress-skin panels are very popular roof decks in the western United States; they are frequently used with glue-lam or heavy timber construction. Their live load capacity is relatively low and consistent with the building code in the area where little snow is expected.

Wood plank construction is frequently observed in centers for religion such as churches, where they form the exposed ceiling and roof deck. Tongue-and-groove planks can be upto 127 mm (5 in.) thick. An attenuation layer to absorb the eventual movement of the planks should be installed before any low-sloped membrane.

Wood fiber plank composed of reconstituted cellulose formed into thick (~75 mm [3 in.]) slabs resembling the cardboard on the back of writing tablets. The slabs are available with square or tongue-and-groove edges and are installed with spikes driven into rafters. "Homasote" is a trade mark that comes to mind.

THERMAL INSULATION

Almost all low-sloped roofing systems use thermal insulation boards either above or, most frequently, below the roofing membrane as part of the over-all system. In most cases thermal insulation performs the following functions:

- Provides a structural bridge across deck corrugations or discontinuities.
- Provides thermal insulation.
- Provides an attenuating layer between the deck and the membrane.
- Provides a reservoir to handle seasonal moisture variation within the system.
- Provides structural support for the roofing membrane and any static or dynamic loads applied to the system.

The structural bridge function is perhaps most important when the insulation is applied to steel decks. The insulation must span the ribs without breaking, and must not be pressed into the ribs under normal construction loads of about 15 kg/m² (300 lb/ft²). Thus, both the flexural strength and the compressive strength of the insulation are important. These properties are not often addressed in the manufacturers' literature, and are sometimes absent in the standards for the insulation. The test for compressive strength reports the peak load, or the load when the sample thickness is reduced by 10 percent. It is therefore quite possible for a 100 mm (4 in.) thick insulation layer to be compressed to 90 mm (3.6 in.) when a load equivalent to its compressive strength is applied. Remember also, that improperly selected roof insulation (most, if not all foams) will be consolidated in areas exposed to repeated impacts, such as just outside entry doors, man hatches, or where rooftop equipment maintenance is frequently necessary. These areas may require reinforcement by rigid pavers or walkway pads. Avoid thin plastic or rubber walkway pads in these areas; they cannot protect the membrane from repeated impact loads.

The thermal resistance property of insulation is measured by several different test methods including the guarded hot plate and rapid k methods. The SI unit for thermal resistance is: metre · degree Kelvin · per watt. In inch–pound units, thermal resistance "R" is square foot · hour · degree F per Btu · inch thickness. The minimum quantity of thermal insulation required is usually specified by the local building code. The optimum quantity of insulation for the roofing system can be calculated using:

$$R_o = \left\{ \left[\frac{480 \cdot M_h \cdot H}{(E_h \cdot B \cdot J)} \right] + \left[\frac{480 \cdot M_e \cdot C}{(E_c \cdot B_c \cdot J)} \right] \right\}^{0.5}$$

where,

- B is the heat content of one heating fuel unit.
- B_c is the heat content of one cooling fuel unit.
- C is the annual cooling degree days (18.3 °C [65 °F] basis) for the location.
- E_h and E_c are the efficiencies of the heating and cooling plant and distribution system as a fraction.
- H is the number of annual heating degree days (18.3 °C [65 °F] basis) for the location.
- J is the cost of the material and installation of insulation with one unit of thermal resistance.
- M_h and M_c are the heating and cooling unit fuel costs respectively, and
- R_o is the optimum thermal resistance.

For a numerical example, for Boston, Massachusetts:

B = (oil) (140,000 Btu) B_c = (electric) (3413 Btu)
C = 661° days E_h = 0.70 (70%)
E_c = 1.00 (100%)   H = 5621° days
J = $0.15 M_h = $1.00
M_c = $0.128

therefore,

R=[(480×1×3621/0.7×140,000×0.15)+(480×0.125×661/1×3413×0.15)]$^{0.5}$
R=[183.54+79.33]$^{0.5}$
R=16.2

The calculation above is for the estimation of the optimum R for the roofing system; this value for R may not be the ideal insulation level for the whole building, because the optimum insulation level for the whole building depends on many more factors than are considered in this book, such as the proportion of window area in the walls, the quantity of wall insulation, the type of glazing, the number of stories, etc.

Attenuation of deck movement is another function of insulation that is not often appreciated. An example of attenuation is when a layer of extruded polystyrene is specified under a membrane such as EPDM or PVC directly on a concrete deck. Here polystyrene acts to absorb any minor deck movements before they can be transmitted to the membrane; it also provides a space to vent any moisture trapped in the system. Polyester fleece – a thin layer of felted polyester – is sometimes used for the same attenuation purpose.

Insulations vary in their capacity to absorb and give up moisture without destroying themselves. This is an important property because the moisture within the roofing system varies as the seasons change. If the insulation can absorb the moisture during damp seasons and give it off during dry seasons, the moisture will have little influence on the other materials in the system. For our purposes, the equilibrium moisture content at 90 percent relative humidity minus the equilibrium moisture at 45 percent relative humidity times the density of the insulation, is called the moisture capacity of the insulation. Wood fiberboard insulation has the highest capacity; lightweight foam insulations the least moisture capacity.

Table 4.2 lists many of the important physical properties of the insulations popular in the year 2001. Note that glass-felt faced gypsum board is included in Table 4.2; it is not strictly an insulation, but it is frequently used within roofing systems as a cover board to strengthen the surface, or under the insulation as a fire stop.

Cellular glass insulation is formed by hydrogen sulfide foaming molten slag. The ASTM International standard that applies is C552 *Standard Specification for Cellular Glass Thermal Insulation*. Blocks of cellular glass are typically faced with kraft paper. Cellular glass is used on rigid decks such as pour-in-place concrete; it is too brittle for use on metal or wood decks. When wet and exposed to freeze-thaw cycles, cellular glass foam disintegrates to black mud. The glass cells under the facer are fairly easily crushed by traffic. Its use is indicated when the insulation is going to be exposed to high (~480 °C [900 °F]) temperatures. There are very few producers.

> *Stinky foam of glass*
> *Is too brittle for most roofs*
> *Support it fully.*

Composite insulation is not a particular type of insulation; it is formed by laminating two or more kinds of insulation, cover boards, or structural planks together in the factory. Insulations frequently laminated include: expanded polystyrene, perlite, isocyanurate foam, and wood fiberboard. Cover boards may be gypsum board, oriented strand board, perlite, plywood, or wood fiberboard. They are promoted to reduce application cost since the roofing contractor can apply two layers of roof insulation at one time. As with the beneficial features of many products, the benefits can frequently be the weaknesses. A single layer of composite insulation has aligned joints. Properly applied two or more layers of insulation have staggered joints that prevent air leaks and thermal loss through the joints. Where the bottom layer is mechanically attached and the cover board is adhered to the bottom layer with hot asphalt, the hot asphalt tends to retard air flow, and the

Table 4.2 Physical properties of low-sloped roofing thermal insulation

Property	Units	Insulation									
		Expanded polystyrene	Extruded polystyrene	Glass fiber-board	Glass foam board	Glass mat-faced gypsum	Perlite board	Phenolic foam	Polyiso-cyanurate foam	Polyurethane foam	Wood fiber-board
Compressive	kPa	69–172	103–862	138	689		241	69–241	110–172	138–414	551
strength	lbf/in²	10–25	15–125	20	100		35	10–35	16–25	20–60	80
Density	kg/m³	12.8–32	22.4–64.1	192	110		177	32–48	27.2–48.1	27.2–48.1	256
	lb/ft³	0.8–2	1.4–4	12	6.8		11	2–3	1.7–3	1.7–3	16
EMC (45% RH)	%	1.9	0.5	0.6	0.15		1.7	6.4	1.7	2	5.4
EMC (90% RH)	%	2	0.8	1.1	0.2		5	23.4	5	6	15
Maximum service temperature	°C	74	74	204	482		93	149	121	121	93
	°F	165	165	400	900		200	300	250	250	200
Moisture capacity	kg/m³	0.0224	0.1300	0.9600	0.0550		5.8548	6.8000	1.2425	1.5060	24.58
	lb/ft³	0.0014	0.0081	0.0335	0.0034		0.3630	0.4250	0.0688	0.0940	1.54
Specific heat	Calorie/(g·°C)	0.27–0.31	0.27–0.31	0.3	0.2		0.25	0.38	0.2–0.25	0.38	0.33
	Btu/(lb °F)	0.27–0.31	0.27–0.31	0.3	0.2		0.25	0.38	0.2–0.25	0.38	0.33
Thermal conductivity	W/(m·K)	0.036	0.029	0.036	0.066		0.055	0.033	0.023	0.086	0.052
	Btu in./(hr·ft²·°F)	0.25	0.20	0.25	0.46		0.38	0.23	0.16	0.16	0.36
Thermal expansion	×10⁻⁶/°C	45–72	45–72	8.4	2.9–8.3		18–23	18–36	54–108	54–108	13–18
	×10⁻⁶/°F	25–40	25–40	4.6	1.6–4.6		10–13	10–20	30–60	30–60	7–10
Thermal resistance	(m·K)/W	27.78	34.48	27.78	15.15		18.18	30.30	43.48	11.63	19.23
	hr·ft²·°F/(Btu·in.)	4.00	5.00	4.00	3.20	1.12	2.63	4.35	6.25	6.25	2.78

cover board assures that the fasteners will not "back out" through the roofing membrane.

Some composites are made with insulations that differ in their thermal expansion coefficients and/or their response to changes in moisture. These composite panels warp when they are exposed. In extreme cases, urethane foam-perlite composites have warped until a section resembles the letter "C."

Expanded polystyrene (EPS) is sometimes called beadboard. It is the plastic used to form hot drink cups from beads of foam, in its most familiar form. Polystyrene beads are heated by steam in a mold to form the insulation board. This product is covered by ASTM International standard C578 *Standard Specification for Preformed, Cellular Polystyrene Thermal Insulation*. This insulation is characterized by its very low density (13–32 kg/m^3 [0.8–2 lb/ft^3]), and high thermal resistance. It is frequently used in cold storage or freezer buildings because of these properties. Thick insulation blocks – often tapered – are adhered together with special adhesives, and perhaps held in place with the aid of wooden pegs. This is a relatively open celled foam that provides a minor barrier to the flow of water vapor and other gasses.

EPS is not indicated:

- on metal decks because of its flammability,
- without a cover board for adhered roofing systems,
- anywhere it is exposed to temperatures greater than ~60 °C (140 °F), such as heat sealing the seams of some single ply membranes, and
- anywhere it will be exposed to strong aromatic solvents such as the benzene or xylene often found in adhesives or cleaning compounds – the beadboard will vanish – it will promptly dissolve.

Extruded polystyrene (XPS) is polystyrene foam extruded to form relatively closed cell boards with a polystyrene skin on the top and bottom faces. It is covered by ASTM International standard C578 *Standard Specification for Preformed, Cellular Polystyrene Insulation*. This is the only insulation that can be used in "inverted" roofing systems – where the insulation is on top of the roofing membrane and covered with ballast. XPS is very resistant to water due to its closed cell construction, but it has styrene foam's weaknesses of low maximum service temperature and poor resistance to solvents. It is not indicated on steel decks, and requires a cover board for any adhered membrane application.

Glass fiberboard insulation currently has a modest 2 percent market share. It is made by bonding glass fiber wool with a kraft paper facer on the top surface. It is covered by ASTM International's specification C726 *Standard Specification for Mineral Fiber Roof Insulation Board*. At one time it was one of the most popular insulations for low-sloped roofing because it could be used directly on steel decks. Glass fiber insulation sales decreased with the advent of the less expensive polyisocyanurate foam insulation. Like foam insulations, glass fiberboard insulation benefits from a cover board to reinforce the surface and improve the impact resistance of the system.

Perlite board insulation is one of the favorite cover boards in roofing systems because of its low cost, fire resistance, and availability. It tends to have a fuzzy surface – making adhesion difficult – and in single 19 mm (¾ in.) thick layers, breaks easily. The latter feature can be overcome using 38 mm (1½ in.) thick layers. Perlite's material standard is ASTM International C728 *Standard Specification for Perlite Thermal Insulation Board*.

Perlite board is made up of expanded perlite aggregate, cellulose fiber (newsprint), and a binder – such as asphalt emulsion. It is sometimes promoted as being water resistant, but it will disintegrate during long-term soaking.

Phenolic foam insulation was manufactured by several organizations during the 1980s and 90s. It has been removed from the market because experience has shown that water flowing from wet phenolic foam severely corrodes steel decking, and there has been some shrinkage noted in fresh insulation. Remediation includes removal of the roofing system, painting or replacing corroded steel decks, and installation of a new roofing system – without phenolic foam insulation.

> *A phenolic foam*
> *Insulation on steel decks*
> *Provides corrosion.*

Polyisocyanurate foam currently has more than a 50 percent market share. It is sometimes called isocyanurate foam, or just plain "iso." The insulation is foamed between platen confined felt facers, and should be aged a while in the warehouse to come to equilibrium before shipping. A cover board is required over the foam for any adhered roofing membrane because of the low cohesive strength of the foam, particularly at the foam–facer interface where the peel strength seldom exceeds 175 N/m (1 lb/in. width). The foam is available in a variety of densities and compressive strengths. Have the compressive strength of the insulation verified by an independent laboratory when high compressive strength is important to the roofing system design; some manufacturers sometimes overstate the compressive strength.

> *Foam insulation*
> *Requires cover boards*
> *To prevent blisters.*

Polyisocyanurate foam insulation is covered by ASTM International's standard C1289 *Standard Specification for Faced Rigid Polyisocyanurate Thermal Insulation Board*. The foam part of the industry is in the process of changing from halogenated blowing agents to agents that are less likely to harm the ozone layer.

> *Polyiso salesmen*
> *Talk about blowing agents*
> *As do most lawyers*

Urethane foam board insulation has been replaced by its polyisocyanurate cousin because polyisocyanurate foam boards can be applied directly to steel decks. There are still installations that contain urethane foam boards. Sliced bun foam boards were faced with asphalt-coated organic felt sheets. Later installations resemble today's isocyanurate foam boards.

Wood fiberboard insulation is the earliest thermal insulation still in use. Fiberboard is a good cover board because it is tough, and has the highest capacity to hold moisture with impunity. It will consolidate when soaked in water. ASTM International's C208 *Standard Specification for Cellulosic Fiber Insulating Board* covers this product.

Fiberboard should be used in 25–38 mm (1–1½ in.) thicknesses. The single-ply 13 mm (½ in.) thickness breaks easily, and the adhesion between the factory laminated plies is sometimes questionable in thicker panels.

Perlite filled asphalt is sometimes used as an insulating fill to insulate and slope concrete decks. The fill is mixed hot on the roof, screeded to elevation, pressed with a heavy garden roller. It is not suitable for steel decks because of the deflection resulting from the application process. The biggest problem is that it is impractical to cover the fill with a roofing membrane fast enough on large jobs. This leaves areas exposed to the weather for some time. When it rains, ponded water floats the installed fill off the deck. This water is almost impossible to remove without removing the fill. Hot asphalt-perlite fill is impractical for roofs with smaller areas because of the relatively high set-up costs. Table 4.3 lists the thermal properties of many materials associated with roofing.

Table 4.3 Thermal properties of some typical building materials

Material	Thermal conductivity		Thermal resistance	
	$W/(m \cdot K)$	$Btu \cdot in./(hr \cdot ft^2 \cdot °F)$	$m \cdot K/W$	$hr \cdot ft^2 \cdot °F/(Btu \cdot in.)$
Asbestos cement shingles	0.69	4.76	1.46	0.21
Asphalt shingles	0.33	2.27	3.05	0.44
Asphalt-perlite fill	0.06	0.40	17.34	2.50
Cellular concrete	0.08	0.55	12.62	1.82
Gravel surfaced BUR	0.44	3.03	2.29	0.33
Gypsum	0.24	1.67	4.16	0.60

Table 4.3 (Continued)

Material	Thermal conductivity		Thermal resistance	
	W/(m · K)	Btu · in./(hr · ft^2 · °F)	m · K/W	hr · ft^2 · °F/(Btu · in.)
Insulating concrete	0.08	0.58	11.93	1.72
Plywood	0.12	0.81	8.60	1.24
Precast concrete	1.80	12.50	0.55	0.08
Reinforced concrete	1.80	12.50	0.55	0.08
Roll roofing	0.96	6.67	1.04	0.15
Slate roofing	2.88	20.00	0.35	0.05
Smooth surfaced BUR	0.60	4.17	1.66	0.24
Steel decks	negligble	negligble	negligble	negligble
Structural cement-fiber	0.07	0.50	13.87	2.00
Wood	0.12	0.84	8.25	1.19
Wood shingles	0.15	1.06	6.52	0.94

Source: Various industry sources.

QUESTIONS

1 Steel deck standards often refer to recommendations by the Factory Mutual Research Corporation. [a] true [b] false.
2 Poured-in-place or stone concrete is considered an insulating and nailable deck. [a] true [b] false.
3 Many steel decks are 38 mm (1½ in.) deep. [a] true [b] false.
4 The ribs on an intermediate rib deck 44 mm (1¾ in.) wide. [a] true [b] false.
5 Poured-in-place concrete decks should be dead level. [a] true [b] false.
6 Concrete decks do not need reinforcing steel. [a] true [b] false.
7 Lightweight concretes are less expensive than conventional stone concrete. [a] true [b] false.
8 Vermiculite concretes must be vented downward. [a] true [b] false.
9 Small clips in steel deck sidelaps provide enough venting for vermiculite concrete. [a] true [b] false.
10 Poured-in-place gypsum deck's biggest disadvantage is that the finished surface is usually dead level. [a] true [b] false.
11 Deck movement can be a problem with metal-banded gypsum planks. [a] true [b] false.
12 Plywood stress skin panels are known for their high load bearing capacity. [a] true [b] false.
13 Be careful on leaking roofs supported by lignin-excelsior planks because the panels might fall off the supports. [a] true [b] false.

14 Insulations just outside doors and man hatches may be consolidated by impacts unless the system is protected. [a] true [b] false.

15 The quantity of thermal insulation used is specified in the Building Code. [a] true [b] false.

16 Attenuation of deck movement is an important insulation function. [a] true [b] false.

17 Wood fiberboard has the highest moisture capacity of any insulation. [a] true [b] false.

18 Cellular glass insulation should be used on steel decks. [a] true [b] false.

19 Composite insulations make better roofs than insulations installed as separate layers. [a] true [b] false.

20 EPS insulation is the same as extruded polystyrene. [a] true [b] false.

21 XPS is the only insulation approved for use in inverted roof assemblies. [a] true [b] false.

22 Polyisocyanurate foam insulation is currently the most popular. [a] true [b] false.

23 All foam insulations require a cover board. [a] true [b] false.

24 Wood fiberboard insulation is a good cover board for other insulations in the system. [a] true [b] false.

25 Hot asphalt-perlite fill insulation can be used on any deck. [a] true [b] false.

5 Steep-sloped roofing systems

Steep-sloped roofing systems started to be used about the time people left the caves and started living in huts with roofs of bark, leaves, or rushes. They quickly learned that a steep-slope to conduct the storm water off the roof was a good thing. Bark, leaves and rushes are still used in many developing countries, and are sometimes used for special effects in other areas. Modern waterproof roofing systems are often installed as insurance under the primitive roofing materials, when they are used today as historic or architectural features. Even with modern steep-sloped systems, waterproof membranes are sometimes used under the water shedding steep-sloped systems, to assure long-term performance.

Recent studies on the service lives of roofing systems show that the steep-sloped systems have service lives much longer than low-sloped systems, confirming again the importance of getting the water off the roof promptly. Table 5.1 shows the estimated average service lives, the fire rating (more about this is explained later in the chapter), and an estimated life cycle cost for steep-sloped systems. These life cycle costs are crude estimates – they do not include maintenance or replacement costs.

Asphalt shingles have long been the most popular steep-sloped roofing material in the United States and Canada. The first asphalt shingles were individual shingles (resembling their bark precursors) made in the early 1900s. The individual shingles were quickly joined together in a broad range of shapes to eventually form the modern "strip" shingle design.

Most of these more modern three-tab strip shingles were made in inch–pound units. Inch–pound three-tab strip shingles are 305 mm (12 in.) high and 914 mm (36 in.) wide. The cutouts separating the three tabs are 127 mm (5 in.) high and 10 mm (⅜ in.) wide. The "exposure" is the height of the cutouts. The area considered as the unexposed area is the upper part of the shingle, roughly 152 mm (6 in.) high by 914 mm (36 in.) wide. The remaining distance, after deducting the exposures of the two overlying shingles, measuring upward from the bottom edge of the shingle, is called the "headlap." The sales square, the gross shingle area required to cover 10.8 m² (100 ft²) of deck, is 26 m² (240 ft²) of sheet (without the cutouts), or 25.6 m² (236.875 ft²) of sheet per net square.

Table 5.1 Steep-sloped roofing mean life, fire classification and life cycle cost

Roofing system	Mean life, y	Fire rating	Installed cost		Life cycle	
			$/m²	$/square	$/(y·m²)	$/(y·ft²)
Concrete tile	50	A	25–28	270–300	0.530	0.057
Heavy asphalt-glass fiber	35	A	16–21	175–225	0.529	0.057
Clay tile	46.7	A	27–37	290–400	0.685	0.074
Asphalt-glass fiber shingles	17.7	A	13–15	145–165	0.786	0.088
Asphalt-organic felt shingles	17.5	C	13–15	145–165	0.800	0.088
Natural slate	60.3	A	74–93	800–1000	1.385	0.129
Architectural metal	26.5	A to C	32–42	350–450	1.386	0.151
Wood shakes (untreated)	12.5	–	23–28	250–300	2.040	0.220
Wood shakes (treated)	12.5	B to C	26–44	280–480	2.800	0.304
Plastic or rubber shingles	~14	C	23–37	250–450	2.143	0.250
Wood-cement shakes	~8	A	28–32	300–350	3.750	0.406
Asbestos-cement slates	31.4	A				

Source: Abstracted from industry sources.

More recently, some "metric shingles" have been offered. They are typically 337 × 1000 mm (13¼ × 39⅜ in.) and have a 143 mm (5⅝ in.) exposure. The basis for all asphalt shingles is the felt or mat used. The felt carries the sheet through the manufacturing and installation processes. During the 1930s, in simple terms, a roll of felt from the felt mill was unwound and fed into a saturator pit filled with hot asphalt. The saturated felt was cooled so the soft asphalt on the surface was sucked into the felt. The saturated felt was then coated on both sides with coating asphalt, an asphalt more viscose than saturating asphalt, and containing inorganic filler such as limestone, slate flour, or equivalent. The coated sheet is top surfaced with roofing granules; the back is surfaced with a mineral parting agent. The surfacing is pressed into the asphalt coating and the sheet is cooled, passed on to the cutter, and the cut shingles are packaged in bundles and palletized for shipment. The usual roofing machine is 400–800 m (¼–½ mile) long.

The dry felt or mat is the source of the shingle's strength. Prior to World War II, roofing felts were so-called "rag felts" composed of felted recycled newsprint, digested wood fibers, and about 8 percent by weight of natural (cotton, wool, or linen) fibers. The use of natural rags had to be discontinued because rag felts were dimensionally unstable (they shrank during asphalt saturation); rayon and other man made fibers that made up a large percentage of the rags sent to the felt mill could not take the heat of saturation with asphalt (the felt fell apart in the saturator – and hand picking the rags to remove the rayon and nylon fabrics at the felt mill was not very effective, was economically costly, and overburdened the personnel available). Many decried the demise of the "good old rag felts" because the newer felts, with the higher wood fiber content, were not as strong in tensile and tear strength as the rag felts, but they were dimensionally more stable and, if properly formed, served well for many years.

Along with the raw materials and personnel availability, pressures were always present to increase production rates of both felt and roofing. This resulted in decreasing the dwell time in the felt mill used to digest the wood fibers (increasing the density of the felt and decreasing the felt's ability to absorb asphalt). Increasing the roofing machine speed reduced the time the felt spent in the saturator and in the drying-in process, and decreased the saturant absorbed. Shingles based on organic felts are still made and preferred to the more modern shingles based on a glass fiber mat or felt because the organic felt based shingles have a higher tear strength than their glass fiber felt based cousins, and are reported to be more flexible in colder climates.

Organic felt based shingles went through an evolution. The first shingles were dipped by hand into hot asphalt, placed on a table with a dusty surface; granules were sprinkled by hand on top of the still hot asphalt and pressed into place with a roller resembling a rolling pin. Maybe 10–20 shingles were made per hour per worker. The shingle production evolved to be performed by the roofing machine described previously. In the 1950s, a felt supplier, a saturator operator, a coater man, a surfacing supplier, a press section operator, a cutter operator, three shingle catchers, a bundle sealer, a palletizer, and a lift truck driver – a total of 12 direct persons serviced such a machine. The sheet speed was about 1 m/s (200 ft/min) that resulted in a production of 1000 shingles per hour per worker, or about 4 kg/s (16 tons/hr of operation). By the early 1960s, automation replaced seven of the workers, and the production rate was 2400 shingles per worker per hour, or about 4.5 kg/s (18 tons/hr of roofing). Today's production rates are even much higher due to the advent of better controls, the advent of glass fiber felts, and additional automation. At one time, the cost of the direct labor per square was less than the cost of the wrapper. In short, the asphalt roofing industry is one of the most efficient manufacturing entities in the United States and Canada.

Manufacturers applied a thicker application of asphalt to the area of the shingle intended to be exposed to the weather to form what was called a "thick butt shingle."

A normal or commodity thick butt shingle had an average mass of 97.5 kg (215 pounds per sales square) or 41.2 kg (90.8 lb/100 ft^2). The unexposed area had a reduced asphalt coating thickness – it was "starved" of asphalt and sometimes had smaller in size granules on the top surface.

This was its undoing. The weather quickly eroded holes in the thinly coated areas under the cutouts. The industry responded by increasing the mean mass from 97.5 to 106.6 kg (215–235 lb) per sales square and provided shingles with a uniform thickness. The current typical asphalt-organic and asphalt-glass fiber felt shingle compositions are listed in Table 5.2.

The seal-tab feature was introduced in the 1950s. Seal-tab shingles boasted a continuous or a dashed line of asphalt or rubber modified asphalt adhesive either on the face of the shingle just above the cutouts, or a dashed line on the back of the shingle near the bottom of the tabs, and a release paper element on the opposed surface. The goal was for the sealant to lightly glue down the bottom edges of the shingle tabs to resist damage from the wind. The adhesive is supposed to use the heat from the sun, and gravity to make the seal. Even a slight seal is enough to resist the force of substantial winds. I observed a wind tunnel test of very slightly sealed shingles that resisted 44.7 m/s (100 miles/hr) winds.

The seal-tab feature was not introduced without a great deal of difficulty. The ideal location for the sealant bars was considered to be the underside of the shingle tabs because:

Table 5.2 Typical composition and physical properties of asphalt shingles

	Asphalt-organic shingles		Asphalt-glass fiber shingles	
	kg/m^2	$lb/100\,ft^2$	kg/m^2	$lb/100\,ft^2$
Composition material				
Granule surfacing	1.59	32.51	1.67	34.28
Top filled coating	1.99	40.80	1.29	26.54
top coating asphalt	0.72	14.67	0.48	9.91
top coating filler	1.28	26.13	0.81	16.63
Dry felt	0.51	10.42	0.07	1.49
saturant	0.81	16.67	–	–
asphalt coating	–	–	0.25	5.11
filler	–	–	0.31	6.35
Back filled coating	0.43	8.87	0.57	11.71
back coating asphalt	0.16	3.33	0.24	4.85
back coating filler	0.27	5.54	0.33	6.86
Back surfacing	0.01	0.19	0.07	1.48
Total mass	5.34	109.46	4.25	86.96
% filler	64		60	
% saturation	160		–	
% saturation efficiency	90		–	
Physical characteristics				
Tear strength, g	2000–3000		1200–1800	
Pull through resistance				

- The adhesive could be applied to the bottom of the shingle the hot adhesive overcoming the bond-breaking nature of the back surfacing.
- The adhesive could find easier adhesion to the rough granule surface in the field (where the abhesive nature of the back surfacing might inhibit adhesion).

But there were also problems with this ideal location. Placing the adhesive on the back side of the shingle tabs meant installation of the paper release strip on either the face or underside of the headlap. This required the shingles to be packaged face-to-back or back-to-back reversed 180°, so the sealant on the bottom of the tabs would be packed against the release paper on or under the headlap. This complicated packaging was solved in several roofing mills, but was never widely accepted in the field, where additional time was required to reorient the shingles. Strippable plastic or release paper strips were tried, but were rejected by the roofers because of the additional time required to strip the tapes prior to installation and the time required to pick up the loose tapes after installation.

The industry finally settled on installing the adhesive on the face of the shingle, just above the cutouts, and installing the release paper on the underside of the shingle. This design has been very effective, when an appropriate adhesive is used.

I have observed roofs in the field where the seal-tab feature did not work because:

- The sealant was glass-like and unable to adhere to the shingle tabs (improper sealant selection).
- There was insufficient sealant present. More than 1 kg of adhesive is required for the shingles needed to cover 10 m^2 of deck (2 lb per sales square). This quantity varies with the shingle design and no consensus standard has been established for the minimum mass of sealant necessary. Where enough sealant is not present, this is due to improper manufacture.
- Dust contaminated the shingle adhesive before the shingle could seal (usually related to cold weather shingle installation).
- The sealant pulled the coating off the shingle. The sealant's adhesion is greater than the adhesion between the coating and the felt. This may mean that the shingle tab was previously manually torn free (for whatever reason), may demonstrate wind damage, or may indicate areas where the shingle saturant did not dry into the felt during manufacture, leaving a weak plane at the felt-coating interface.

Excessive seal-tab adhesion is often related to what has been called "thermal splitting" in asphalt-glass fiber strip shingles. This horizontal and vertical splitting of shingle tabs is only observed in glass fiber felt shingles with the tabs firmly adhered. Splits of this type have never been reported in asphalt-organic felt roofs, or in asphalt-glass fiber felt roofs with stronger glass felts, or with unadhered shingle tabs. Two splitting mechanisms have been postulated. One hypothesis is that shrinkage due to normal temperature declines induces forces at stress concentrations to tear or rupture the shingles. Another hypothesis suggests that wind flutter causes the glass fibers to slide out of the asphalt matrix at points of stress concentration to form the splits. These hypotheses do not contradict each other; they merely suggest different forces as the origin of the splitting. It does not change the fact that whatever the forces involved, insufficient strength was due to the glass fiber felt selected by the shingles' manufacturer.

The asphalt saturant in the organic felt retards the absorption of water, may provide some light oils to preserve the stiffer asphalt coating, and also supplies the fuel during the fire resistance tests typically performed on roofing materials. These tests are often considered "performance tests" even though no direct correlation has ever been established between performance in these tests and fires. Despite the lack of correlation, these tests are performed regularly, and are considered important by most building code authors. The tests and the requirements for passing each fire classification (Class A, B, or C) are listed in Table 5.3. Of course it is possible that the candidate system will not meet the requirements of any class – and not be rated. Unrated roofs may provide no protection from fire.

Table 5.3 Fire test interpretations; Classes A, B, and C

Rating	Class C	Class B	Class A
Intermittant flame	704±28 °C (1300±50 °F) flame 3 cycles, 60 s on, 120 s off	760±28 °C (1400±50 °F) flame 8 cycles, 120 s on, 120 s off	760±28 °C (1400±50 °F) flame 15 cycles, 120 s on, 120 s off
Burning brand	5–37×37×20 mm (1½×1½×25/32 in.) pine brands at 60 to 120 s intervals	2–147×147× 57 mm (6×6×2¼ in.) Douglas Fir brands	1–305×305×57 mm (12×12×2¼ in.) Douglas Fir brands
Spread of flame	704±28 °C (1300±50 °F) flame =/< 4 m (13 ft) spread in 4 min	760±28 °C (1400±50 °F) flame =/< 2.4 m (8 ft) in 10 min	760±28 °C (1400±50 °F) flame =/< 1.8 m (6 ft) in 10 min
Flying brand	704±28 °C (1300±50 °F) flame for 4 min	760±28 °C (1400±50 °F) flame for 10 min	760±28 °C (1400±50 °F) flame for 10 min
Rain	12 one-week cycles – 96 h of water at 1.8 mm/h (0.07 in./h) – 72 h drying at 60 °C (140 °F) then subjected to the intermittant flame, burning brand, and flying brand test		
Outdoor weathering	one, two, three, five, and ten years subjected to the intermittant flame, burning brand, and flying brand test after each exposure period		

Source: Abstracted from ASTM E108 and other industry sources.

Class C is the lowest fire classification. Class C roof coverings are effective against light fire exposures. Under such exposures, roof coverings of this class are not readily flammable, afford a measurable degree of fire protection to the roof deck, do not slip from position and pose no flying brand hazard. To pass, the wood sample deck must not ignite and burn severely enough to be destroyed, unless the fire is extinguished. Asphalt-organic shingles are typically classified as Class C. Once I supervised a special run of asphalt-organic felt shingles manufactured without any saturant in the felt to investigate the function of asphalt saturant during the fire tests. These saturant-less shingles passed all the Class A fire tests, but these shingles were not suitable for exposure to the weather. Water would quickly penetrate the unsaturated felt and blister the coating and surfacing off the felt, if such shingles were to be exposed to the weather.

Class B exists, but is very seldom used. Some wood shakes, treated for fire resistance, have a Class B designation.

Class A is the highest fire resistance classification. It is held by most of the modern steep-sloped roofing systems.

Asphalt-glass fiber felt based shingles have largely replaced the asphalt-organic felt shingles. Asphalt-glass fiber shingles simplify the earlier manufacturing process. The glass fiber felt must be preheated (to drive off any water or unreacted glass binder present), and is sent directly to the coater. The rest of the process is fundamentally the same as the methods used for organic felt shingles. Note the absence of the saturator. The absence of asphalt saturant enables the asphalt-glass fiber felts to pass all the Class A fire resistance tests.

Asphalt-glass fiber shingles have lower tear strengths and tend to be more brittle than asphalt-organic shingles. Typical glass based shingles cannot survive even a mild wind storm without an effective seal-tab feature. For this reason, we require the roofer to examine the adhesion of the tabs about 30 days after the shingles are installed, and require the shingler to hand-seal any unsealed tabs.

Filler, or mineral stabilizer, as many manufacturers like to refer to it, is finely powdered mineral that thickens the asphalt coating. Many materials have been used in the past including sand, green slate flour, limestone, micaceous fractions, diatomatious earth and mine tailings. Generally, the quantity of filler that can be incorporated into asphalt increases with the mean diameter of the particles, reducing the cost of the filled coating. But there is a practical limit to the filler size that can be used effectively. Using coarse sand as filler, with its low surface area, results in an unacceptably brittle shingle coating.

A few fillers do deserve the title of stabilizers because, in the right quantities, they increase the durability of the asphalt coating. Fillers with plate-like particles like slate flour are particularly effective in prolonging the life of asphalt coating, but many of the mines have been closed based on the recommendations of the Environmental Protection Agency because the miners have too much exposure to dust which can result in silicosis. Most of today's manufacturers use some variant of limestone as their filler. In organic felt based shingles, filler is usually about 50 mass percent of the filled coating. Filler is about 60 mass percent of the filled coating in glass felt based shingles.

Granule surfacing provides the primary protection against solar degradation. Properly adhered, these opaque minerals reject sunlight, assist fire resistance, and provide color to the shingles. Over time, various minerals have been tested as roofing granules, including pigmented and bare sand, blast furnace slag, apatite, nephaline cyanate and basalt. Of these, only the most opaque minerals make suitable granules. A dense mineral such as apatite has opacity of about 95 percent, but granules made from apatite fall like rain off shingles after about two years of exposure.

Asphalt shingles must retain their granules in order to survive. I once assisted in an experiment to test the function of granules. We manufactured roofing shingles with and without granule surfacing and exposed them to the weather. The shingles without surfacing failed within three months, while the shingles with surfacing behaved normally.

Asphalt shingles are still evolving. Laminated shingles (shingles with several layers glued together in the factory) promise to provide more protection than the simple three-tab strip shingle and are increasing in popularity. Each layer must be properly manufactured and correctly installed in order to get the increased protection promised. I am concerned that some manufacturers may elect to use less asphalt coating, too much filler, or weaker felts in the name of "cost improvement". History shows:

Manufacturers
Cost improve all their products
Until they don't work!

The installation of these laminated shingles is often complicated, and failure to follow the instructions on each package can lead to less than ideal performance and homeowner dissatisfaction.

More recently, very flexible shingles have been manufactured using SBS polymer-modified asphalt for the coating. A typical sample shingle, shipped to you in a mailing tube, falls flat on the table when it is removed from the tube. It is hoped that this greater flexibility and improved granule adhesion will provide a substantially longer service life, but we must always await the test of time; the exposure history developed, because there is no single test or any battery of tests that accurately predicts future performance of the materials.

Architectural metal systems are finding increasing utility as the systems improve and as more suppliers enter the market. Copper was probably the earliest roofing metal and is still valued highly for its durability and the green-to-black patina that develops with age on the surface is highly prized.

Copper can be corroded by acid rains and emissions from coal- or oil-burning power plants. Standing and batten seams are very popular methods for joining copper sheets. Flat seams, used at the top of domes, or where drainage must cross the joints in the copper, are also used – but be careful. Flat copper seams can be made watertight, but only with great difficulty. Be sure to provide properly spaced expansion joints in flat seam roofs; remember that a carefully soldered flat seam roof segment behaves like a single piece of copper, and avoid points of stress concentration such as re-entrant corners.

Cleaning and tinning the copper in the joint area just before it is soldered with heavy irons can maximize the strength of the soldered joints between copper sheets. Lead is another old time roofing material. Its use on roofs may predate the use of copper. A story in the industry, which may be apocryphal, talks about the caretaker for the great cathedral in Cologne, Germany. He is reported to have said that the original lead roof lasted 300 years; the replacement only lasted 150 years – they're not making things the way they use to. Without stress, lead is very durable – as evidenced by the waterproof lead lining in the Roman bath at Bath, England.

Lead creeps under load, and therefore should not be used where it is under constant load. It is often incorrectly specified for "lead wedges" to hold flashing or railing posts in place. The lead in these applications starts to loosen by creep as soon as the initial placement load ceases. If, for example, the flashing is under a constant load, the wedges will fall out and the flashing will be displaced to admit water. Use screwed-in anchors to fasten flashing into reglets – if you must use reglets (see Chapter 6 on flashing). Lead should not be used in direct contact with mortar or other sources of free alkalinity; it will corrode away.

Lead is frequently used to coat steel and aluminum to improve the weather resistance they exhibit. Lead-coated copper is particularly useful and preferred to copper where storm water drains over flashing and onto masonry. Water draining across copper flashing will stain the masonry below with green streaks.

Galvanized or zinc-coated steel is frequently used for flashing; its use is not recommended near seacoasts and the associated exposure to salt water. Currently, the specifications for high quality steel decks for built-up roofing require galvanized metal.

Structural sheet metal such as corrugated steel or aluminum sheet roofing is gaining popularity. The steel sheets must be coated with some type of weather resistant paint or coating. The aluminum sheets are often coated to get colors desired by the marketplace. These coatings can be scraped or cut during shipping and installation to expose the base metal. Subsequent corrosion often eats under the coating, permitting it to peel away from the damaged or cut area, which leads to owner dissatisfaction and disputes.

Generally it is wise to keep corrugated aluminum and steel away from the seashore and seagulls. Corrugated metal roofs resemble large salt or pepper shakers when viewed from the interior, due to the salt content of seagull droppings.

Points of fixity, such as fasteners and clips, penetrations such as HVAC ducts or skylights that span several panels, all provide great difficulties with metal roofing systems. These difficulties can all be traced to the thermal movement of the metal roofing that must be accommodated (it cannot be stopped as some people devoutly wish). Another problem is that we neither have the long experience needed to devise and properly install the intricate flashing details required, nor the skill to correctly line up the metal hold-down cleats so that they permit the expansion and contraction of the corrugated panels, without causing them to buckle. Neoprene or EPDM washers have been tried to seal fasteners that are exposed to the weather – with limited success. The best plan:

- Don't use face fasteners.
- Use single metal sheets from ridge to eave.
- Use a watertight, flexible expansion joint at the ridge.
- Severely limit the number of deck penetrations – keep them small and flexible.

Remember not to locate walkways or roadways under the eaves of a metal roof in northern climates. Accumulated snow and ice suddenly cascading off metal roofs has crushed cars, and periodically closes 54th Street in New York.

Most of the people involved with metal roofing are sheet metal workers (if we are fortunate – ones with roofing experience), but experienced or not – they seem to have the same cure for every problem: "Get out the tube of sealant." This band-aid approach is fine for a temporary fix, but the adhesion and cohesion of sealants are short-term properties and they do not solve mechanical problems; these require sealant free designs.

Werner Gumpertz sez:
"All sealants are hole-fillers
Not waterproofers"

Table 5.4 lists the estimated expansion and contraction coefficients for a number of materials. Most of the plastics have higher coefficients than the metals, but these soft materials can absorb the thermal movement much easier than the stiffer metals. Soft or not, these data remind us to design with care and consideration for the movement in building materials caused by thermal cycles. One conference speaker (and I am sorry that I do not know his name) said: "A building moves like a belly dancer – only slowly."

Natural state is still used on prestigious buildings. Its most difficult feature is that the designer must design the structure for the not inconsiderable weight of the slates. Other problems include obtaining good quality slates (some of the imported slates leave a lot to be desired), and the availability

Table 5.4 Estimated thermal expansion of some building materials

Material	~Expansion coefficient		Unrestrained movement	
	/°C($\times 10^{-6}$)	/°F($\times 10^{-6}$)	mm/m/50 °C	in./100 ft/100 °F
Ice[a]	−113	−63	−5.65	−7.56
Quartz	0.59	0.33	0.03	0.04
Pyrex glass	4	2.2	0.20	0.26
Fired clay	5.5	3	0.28	0.36
Limestone	7.5	4.2	0.38	0.50
Asbestos Cement	7.5	4.2	0.38	0.50
Marble	8	4.4	0.40	0.53
Slate	8	4.4	0.40	0.53
Glass	9	5	0.45	0.60
Granite	9	5	0.45	0.60
Concrete and mortar	11	6.1	0.55	0.73
Mild steel	11	6.1	0.55	0.73
Iron	12	6.7	0.60	0.80
Steel	13	7.2	0.65	0.86
Stainless steel (austenitic)	17	9.4	0.85	1.13
Copper	17	9.4	0.85	1.13
Brass	19	11	0.95	1.32
Aluminum	24	13	1.20	1.56
Asphalt-glass shingles	25	14	1.25	1.68
Lead	29	16	1.45	1.92
Zinc	31	17	1.55	2.04
Poly [vinyl chloride]	50	28	2.50	3.36
Polycarbonate	70	39	3.50	4.68
TPO membrane	75	41.7	3.75	5.00
EPDM	92	51	4.60	6.12
PVC membrane	98	54	4.90	6.48
Asphalt BUR	213	118	10.65	14.16
Coal-tar BUR	290	161	14.50	19.32
APP PMA membrane	296	164	14.80	19.68
SBS PMA membrane	336	187	16.80	22.44

Source: Assembled from industry sources.

Note
a 0 to −10 °C.

of qualified slaters who will sound each slate (strike it with a hammer to see if it rings – or is sound) before it is installed.

Check samples of the slate to be sure that they comply with ASTM International's Standard C 407 and listen to your experienced applicator. Unfortunately, retirement and a failure to train new slaters limits the number of qualified workmen. Investigate proposed applicators as you would a prospective employee; experience and care count. Be sure to use stainless steel or bronze nails. It makes little sense to install long-lived slates with short-lived fasteners.

Concrete tile has the lowest estimated life cycle cost in Table 5.1. This cost does not include the periodic cleaning and recoating these tiles require. Concrete tiles are frequently used in Florida. The typical construction is:

- A plywood or oriented strand board layer nailed to the rafters.
- A layer of granule surfaced, asphalt-coated glass fiber felt (the "undertile") nailed to the deck, with hot asphalt sealed side and end laps.
- Concrete tiles adhered to the deck with globs of sand–cement mortar – or more recently, foamed-in-place low-rise urethane foam.

Here, the undertile is the waterproofing element. Some of the problems observed include:

- Tile displacement down slope and in high winds due to questionable adhesion between the tile and the mounds of mortar.
- Tile breakage from traffic due to inadequate tile design or formulation.

Use of one of the urethane adhesives designed for the application probably will eliminate the adhesion and sliding problems. Care in product selection is needed to avoid the breakage problem. Concrete tiles are not suitable for locations that experience freeze-thaw because of the high water absorption they exhibit.

Ceramic tile is the design selection in many locations; flat tiles in many areas; curved "Spanish" tiles in locations that relate to a historic Spanish culture. The ancient art of hanging tiles on spaced wooden slats is seldom observed. Many of these tiles are installed in a similar fashion to concrete tiles; others require mechanical fasteners. Ceramic tile problems are similar to those listed for concrete tile. Be careful; not all ceramic tiles are suitable for use in areas that experience freeze-thaw. Review all literature and check test results.

Wood shingles or shakes are very popular in the western and Pacific coast regions of the United States. Some people rave about their beauty. Beauty may be in the eye of the beholder, but wood shake or shingle roofs look to me like a conflagration waiting to take place. Many lower cost and higher fire resistant systems are available and should be used.

Fiber-cement tiles or slates are relative newcomers to the market. Currently, for all practical purposes, outlawed asbestos-cement shingles have been replaced by wood fiber-cement products of various types. These candidate products were made by many organizations, at geographically separate locations. The producers apparently believed that all that was required was to replace the asbestos fibers with wood fibers in several forms including newsprint, wood splinters, or wood flakes. All of the producers offered 25–50 year warranties, and almost all of them have gone into bankruptcy when the shingles or shakes started to fail. Despite the large investments in plant, people, and technology, they failed to appreciate a fundamental fact: you cannot make a durable product for outdoor exposure out of Portland cement and wood fiber – or any other additive that absorbs water.

> *Phony roofing slates*
> *Of cement and wood fiber*
> *Really suck – water.*

The slate or shake roofing elements fail in several patterns including exfoliation, delamination, warping, cracking, and turning to mush. These very different modes of failure probably obscured the fundamental water absorption problem. I'm quite sure that each producer felt the answer to his problem was right around the corner, but this kind of wishful thinking is akin to desiring the repeal of the law of gravity – a very unlikely occurrence.

In Table 5.1, I estimate the average life of a wood fiber-cement shingle or shake as 8 years. This is quite generous based on my experience. How then did the manufacturers warrant the performance of the products for 25–50 years? One answer heard was: "They don't expect to be in business in 50 years." The situation is much more complex.

There is no test, or body of tests, that can predict the service life of a roofing system. Regardless of how we define failure, the broad variation in climates, exposures, and differing failure mechanisms, make accurately predicting the future highly improbable – perhaps that is why they stoned seers in olden times. To make matters worse, the rate of innovation is increasing, so that there are almost no products that have lasted for the duration of the warranty. Given that the service life of a product can not be determined; wouldn't we be better off without warranties? In that case maybe the designers would take more careful looks at the systems they are specifying. Or alternatively, given that the manufacturers have caused the owners to change their position in equity based upon a baseless promise; doesn't this seem like fraud? Perhaps when one person does it; he's a crook. If a whole industry does it; it's puffery – and should not be believed by any thinking person. The last would seem the safest course – ignore warranties.

QUESTIONS

1   Steep-sloped roofing systems have longer service lives than low-sloped systems. [a] true [b] false.
2   The exposure on conventional three-tab strip shingles is [a] 4 inches [b] 6 inches [c] 5 inches.
3   The headlap on three-tab strip shingles in inch–pound units is [a] 1 inch [b] 2 inches [c] 3 inches.
4   "Rag felts" contained about 8 percent natural rags. [a] true [b] false.
5   In the 1950s, a roofing machine was run by [a] 2 people [b] 5 people [c] 10 people [d] 12 people.

6 Modern asphalt-organic strip shingles weigh about [a] 150 pounds per sales square [b] 235 pounds per sales square [c] 215 pounds per sales square.

7 The seal-tab feature is essential to the performance of asphalt-glass shingles. [a] true [b] false.

8 The seal-tab feature may not work due to [a] improper adhesive [b] insufficient adhesive [c] dust contamination [d] all of the previous reasons.

9 Thermal splitting is related to excessive seal-tab adhesion. [a] true [b] false.

10 Asphalt-glass shingles typically have a Class A fire rating. [a] true [b] false.

11 A Class B fire rating is widely used. [a] true [b] false.

12 Asphalt-glass shingles have a higher tearing strength than asphalt-organic shingles because glass is much stronger than wood. [a] true [b] false.

13 The proportion of filler in an asphalt coating depends on the size of the filler. [a] true [b] false.

14 Granule surfacing is the primary protection against solar degradation. [a] true [b] false.

15 SBS polymer modified asphalt-glass shingles are stiffer and stronger than shingles made with unmodified asphalt. [a] true [b] false.

16 Flat seam copper roofing segments respond to temperature changes like a single piece of copper. [a] true [b] false.

17 Lead makes excellent wedges to pin flashing into reglets. [a] true [b] false.

18 Use metal roofing whenever the roof has many penetrations. [a] true [b] false.

19 Sealants are hole fillers, not waterproofers. [a] true [b] false.

20 Tiles require a waterproofing membrane under them to perform properly. [a] true [b] false.

21 Wood shake roofs should be avoided in high fire risk exposures. [a] true [b] false.

22 Fiber-cement tiles and slate roofs should be avoided because they quickly deteriorate in the weather. [a] true [b] false.

23 Carefully selected laboratory tests accurately predict the performance of roofing systems. [a] true [b] false.

24 The expected life of a roofing system is shown by the warranty. [a] true [b] false.

6 Flashing

For our purposes let us call the special roofing features "flashing" at the intersections of:

- differing roofing systems,
- differing planes in the same system,
- roofing and penetrations,
- roofing and rising walls, and
- roofing and its perimeters.

Flashing is perhaps the most critical area of any roofing system because it is the area where stress is likely to be concentrated; it is the area that requires the greatest time, attention, and skill of the designer as well as the roofing mechanic. It is also the first place to look for any problems, including wear and dislocations.

Some of the most frequently observed errors in flashing systems are:

- Failure to match the flashing and roofing systems. This most frequently takes place for example, when a built-up roofing system is switched to a single ply system (or vice-versa), and no one remembers to change the original flashing system design. This leaves the flashing design up to the mechanic who may or may not be familiar with the requirements of the new system.
- Failure to match or mate flashing systems where they must intersect. For example, providing a clear vertical section through the wall to roof detail, but ignoring the flashing on the adjoining wall.
- Failure to connect wall and roof structural expansion joints (the building movement will connect them, and the resulting tear will admit water and leak).
- Failure to detail the place where the flashing detail ceases, such as at a roof rake running out into a roof plane (a very difficult detail), or such as the flashing at the ends and intersection of the roof edge and wall flashing.
- Failure to carry expansion joint details through the perimeter flashing.

- Failure to carefully detail the work. This is shown by the lack of details, or the reliance on the manufacturer's recommended details (most manufacturers' recommendations do not show a drain detail), or requiring the contractor to provide shop drawings for all flashing details.
- Providing impractical or unbuildable details, or details that do not consider the job requirements to keep the building watertight.

There are a few frequently observed flashing details that should be avoided when possible including, but not limited to:

- *Pitch pockets* – these are flanged bottom and topless metal boxes that surround the roof penetrating elements such as conduits, pipes, and fence railing supports. The box about the penetration is usually partly filled with mortar and topped with either low melt asphalt, coal tar, or pourable sealant. The fundamental problem with pitch pockets is the inability of the topping to keep the detail watertight. Bitumens and sealants require constant maintenance; they crack and separate to admit water. As usual, designers who rely on the owner's maintenance to keep their design watertight, are doomed to disappointment.

> *Unfilled pitch pockets*
> *That are not ponding water*
> *Demonstrate leakage.*

- *Face reglets* – are stiff bars fastening the top edge of a base or counter-flashing. The top edge of the bar is shaped to receive a bead of sealant to waterproof the joint to the wall. As stated earlier, a sealant bond is at best temporary, and subject to leakage. The butting joints of the bar may tear the flashing by the response to thermal cycling and the exposed fasteners of the bar may provide a leakage path. Cracks or any irregularities in the wall surface provide unsealed paths for leakage. Face reglets are particularly ineffective on walls of brick, concrete block, split faced block, and vertical wood siding.

> *Face reglet details*
> *Direct rainwater inside*
> *The flashing detail.*

- *Reglets* – are slots in a surface, shaped to receive, secure, and waterproof the top flashing edge. Reglets are sometimes a necessity – such as when a flashing must be terminated on an existing precast concrete wall in re-roofing or when used to secure metal step flashing about a brick chimney. In new roofing, reglets should be avoided by using through wall flashing. As with face reglets, reglets should be avoided on rough or cracked surfaces, because the depth of the reglet may have to be impractically deep to account for the surface irregularities. In remedial roofing work, new reglets sometimes show the old flashing and weep holes were covered over.

- *Incomplete cap flashing* on top of a parapet guarantees leakage. Here, the designer wants to cap the interior face of the parapet, but does not want to see the metal flashing on the outside of the parapet. The usual way this is attempted is to provide for a metal cap flashing that extends half way across and dives into a horizontal reglet on the top of the parapet. The metal forms a funnel to direct water into the reglet.

> *Half a cap flashing*
> *Is like being half pregnant.*
> *Totally useless.*

Flashing design parameters vary with the type of roofing or waterproofing system. In waterproofing, the flashing is usually installed before the general field of the deck is covered. I consider inverted roofing, where the extruded polystyrene foam insulation is installed on top of the roofing membrane, as waterproofing.

> *Inverted roofing*
> *Is really waterproofing.*
> *Install flashing first.*

Single ply roofing must be mechanically fastened at each penetration and perimeter, at the same elevation as the general roofing surface. The alternative, running the roofing sheet up a wall and attaching the sheet along the top edge, usually results in the sheet tenting off the wall, pulling off the fasteners, and inviting substantial leakage.

Each single ply system has its own special requirements and flashing needs. These are too varied to discuss in this book. Be sure to study the manufacturer's details carefully, and have the manufacturer's technical representative or a competent consultant specializing in roofing, review the details before the job is sent out for bid.

> *Detailed peer review*
> *Before the design is bid*
> *May save big money!*

Consulting an experienced roofing contractor about build ability can frequently be useful, but remember: the roofing designer has the responsibility to select the roofing system and details to be used and should not abrogate that responsibility to anyone.

In the United States, the federal government currently requires the designer to be generic in the specifications – allegedly to promote competition. This gives rise to specifications that require the contractor to provide: "…a 20 year built-up, single ply, or polyurethane foam roof – or equal." While the goal may be meritorious, it also removes design responsibility from the designer. This makes no sense at all; the designer should have the obligation to select appropriate materials and systems. Too frequently disasters have been caused by generic specifications or changes made to specifications by well-meaning politicians, owner's representatives, or others. Of course, some of the changes are not motivated by what is best for the job, but a prudent designer must be willing to listen to anyone – and then make the correct, clear, and informed decision.

Use propriety systems only where appropriate and necessary. Suppliers tend to enter a price competition when generic materials are specified. That does not mean it is appropriate to specify: "built-up roofing, single ply, or equal." Specify a built-up system by any of several manufacturers – if that is what the job demands. There are enough PVC, APP-PMB, SBS-PMB, and EPDM suppliers to provide competition. Do not specify: "built-up, EPDM, PVC, or equal," because the flashing systems for bituminous, rubber, and thermoplastic systems differ significantly.

Designers design.
It is hoped contractors build.
Don't get them confused.

Built-up and polymer modified bitumen roofing generally require similar details that involve elevating each penetration, except drains, over the general roofing surface, and gentle transitions in plane. The following 17 details have inch–pound dimensions. These can easily be converted to SI millimetres by multiplying the inch values by 25.4. Lumber is detailed in nominal values. "2 by" lumber is 38 mm (1½ in.) thick.

Interior roof drains (as shown in Detail 6.1) should be sumped with tapered insulation so that ice at the drain clears before ice on the general roof elevation. The sump also allows for the increased thickness of the flashing – without damming the water flow. Typically, all the roofing plies are cut off at the inner diameter of the drain. A sheet of lead flashing is carefully shaped into the sump, set in a full bed of asphalt flashing cement, and a hole is cut approximately 25 mm (1 in.) inside the edge of the roofing. The edge of the lead at the drain is peaned down into the leader to cover the edge of the roofing. The drain clamping ring secures the lead and the felt plies. Two sheets of asphalt-glass fabric set in flashing cement seal the lead sheet to the roofing membrane. In polymer modified asphalt roofing, the top ply of the drain flashing is a granule-coated cap sheet.

A typical *sanitary vent pipe detail* is shown in Detail 6.2. Note that the edge of the roofing is elevated and that wood nailers are provided to fasten the edge of the roofing and the flange of the metal pipe flashing. The metal pipe flashing can be copper, stainless steel, or lead-coated copper. The

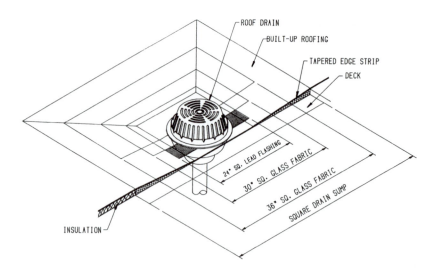

Detail 6.1 Roof drain.

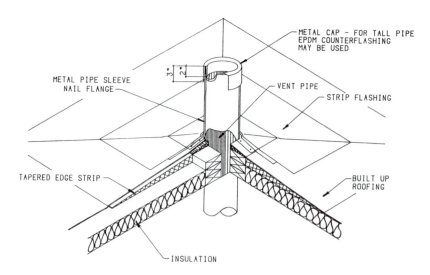

Detail 6.2 Vent pipe penetration (metal cap).

metal cap should be tack welded or soldered in place. In southern climates a 6×6 mm (¼ × ¼ in.) screen may be soldered to exclude rocks and other debris from vandals or pranksters. Do not use screening in northern climates; they will freeze over if used.

The top of a vent pipe flashing or other pipe penetrating flashing can be sealed off with a bond breaking and EPDM tapes as shown in Detail 6.3.

Detail 6.4 shows one way to flash a *multiple pipe penetration*. This is a wood box, capped with metal. For a hot pipe, such as a chimney penetration – omit the plywood top and stuff the box with fiber glass batt insulation.

A typical *curb flashing detail* is shown in Detail 6.5. Metal counterflashing, similar to that in Detail 6.6 is not shown, but must be used to cover the nails at the top of the base flashing. *Duct flashing detail* is shown in Detail 6.6. The counterflashing here shall be sheet metal screwed or pop riveted in place on the duct with fasteners 100–150 mm (4–6 in.) on centers.

Roof mounted equipment should be minimized by installing equipment such as HVAC (heating, ventilating, and air conditioning) equipment on pads off the roof or in machine rooms where the equipment can be properly maintained and protected. Despite good advice, some clients still want to install condensers and other equipment on the roof. Whenever this takes place, install the equipment on watertight, completely protected blocking such as shown in the *equipment support blocking detail*, Detail 6.7. Do not install equipment on wood blocking resting on the surface of the roof. The weight and vibration of the equipment will cut or erode through the membrane eventually. Alternatively, support the equipment on a frame

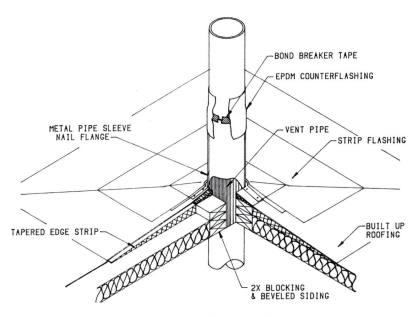

Detail 6.3 Vent pipe penetration (EPDM counterflashing).

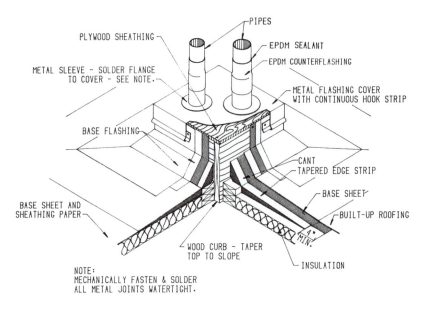

PIPES
PLYWOOD SHEATHING
EPDM SEALANT
METAL SLEEVE - SOLDER FLANGE
TO COVER - SEE NOTE.
EPDM COUNTERFLASHING
METAL FLASHING COVER
WITH CONTINUOUS HOOK STRIP
BASE FLASHING
CANT
TAPERED EDGE STRIP
BASE SHEET
BASE SHEET AND
SHEATHING PAPER
BUILT-UP ROOFING
4"
MIN.
WOOD CURB - TAPER
TOP TO SLOPE
INSULATION
NOTE:
MECHANICALLY FASTEN & SOLDER
ALL METAL JOINTS WATERTIGHT.

Detail 6.4 Multiple pipe penetration.

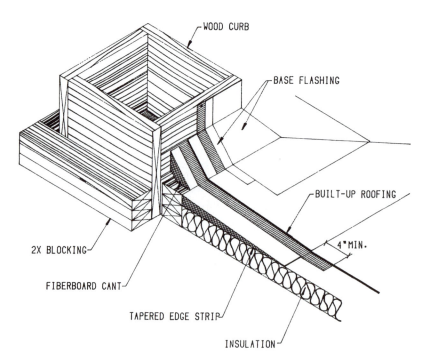

WOOD CURB
BASE FLASHING
BUILT-UP ROOFING
4"MIN.
2X BLOCKING
FIBERBOARD CANT
TAPERED EDGE STRIP
INSULATION

Detail 6.5 Typical curb flashing.

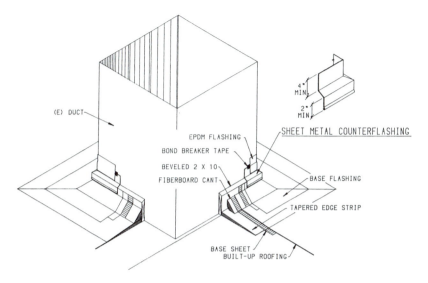

(E) DUCT

EPDM FLASHING
BOND BREAKER TAPE
BEVELED 2 X 10
FIBERBOARD CANT

SHEET METAL COUNTERFLASHING

4" MIN

2" MIN

BASE FLASHING

TAPERED EDGE STRIP

BASE SHEET
BUILT-UP ROOFING

Detail 6.6 Metal duct penetration.

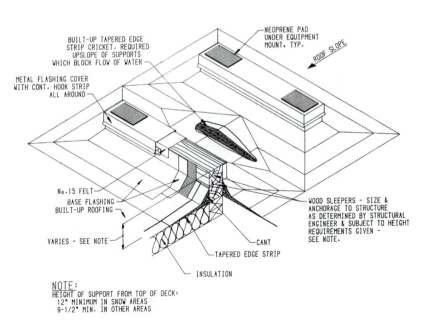

BUILT-UP TAPERED EDGE
STRIP CRICKET. REQUIRED
UPSLOPE OF SUPPORTS
WHICH BLOCK FLOW OF WATER

NEOPRENE PAD
UNDER EQUIPMENT
MOUNT, TYP.

ROOF SLOPE

METAL FLASHING COVER
WITH CONT. HOOK STRIP
ALL AROUND

No. 15 FELT
BASE FLASHING
BUILT-UP ROOFING

VARIES - SEE NOTE

WOOD SLEEPERS - SIZE &
ANCHORAGE TO STRUCTURE
AS DETERMINED BY STRUCTURAL
ENGINEER & SUBJECT TO HEIGHT
REQUIREMENTS GIVEN -
SEE NOTE.

CANT
TAPERED EDGE STRIP

INSULATION

NOTE:
HEIGHT OF SUPPORT FROM TOP OF DECK:
  12" MINIMUM IN SNOW AREAS
  9-1/2" MIN. IN OTHER AREAS

Detail 6.7 Equipment support blocking.

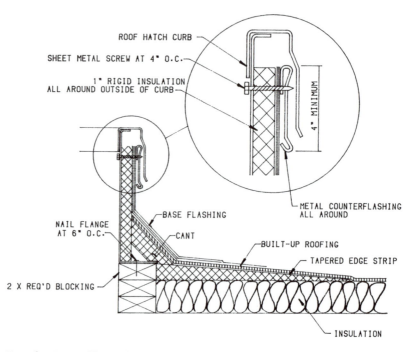

ROOF HATCH CURB

SHEET METAL SCREW AT 4" O.C.

1" RIGID INSULATION
ALL AROUND OUTSIDE OF CURB

4" MINIMUM

METAL COUNTERFLASHING
ALL AROUND

BASE FLASHING

NAIL FLANGE
AT 6" O.C.

CANT

BUILT-UP ROOFING

TAPERED EDGE STRIP

2 X REQ'D BLOCKING

INSULATION

Detail 6.8 Roof hatch flashing.

with round metal legs (for easy flashing as in Detail 6.3) of sufficient height so that roofers, without being dwarfs, can easily work under the frame. Be sure to check any elevated equipment for stability under wind, and where appropriate – earthquake and drifting snow loads.

The flashing at roofing man hatches can be a problem. Currently, these are prefabricated of heavy gauge steel, and have room for 25 mm (1 in.) of insulation on the walls of the curb. It is very difficult to get the top of the flashing under the roof hatch-curb cap that is usually welded in place. One effective solution is to fasten the flashing and a metal counterflashing with sheet metal screws installed from the interior of the hatch. On the side of the curb across from the ladder, install a longer piece of 0.071 mm (22 gauge) metal counterflashing to cover the outside face of the flashing on the curb and cant strip. This is to prevent the unintentional hole kicked in the flashing by a boot worn by a worker. Most Canadian flashing details include metal counterflashing that extends almost to the surface of the roof.

Parapets, or perimeter walls about roofs have many uses. They tend to:

• make buildings appear taller,
• even out roof elevation,
• hide roof top equipment,

- protect adjacent buildings from wind blown gravel, and
- prevent firefighters from walking off a smoke filled roof (parapets or perimeter fences are required in New York City for this reason).

The disadvantages of parapets are that they tend to be:

- not strong enough to resist a seismic event (for brick or CMU masonry parapets),
- dams to pond storm water or water added during firefighting, should the drainage system become plugged – which may lead to the collapse of the building,
- a source of water intrusion from condensation in unsealed cavity walls, or from storm water through optimistic (I hope it works) flashing, and
- they tend to move differently than the insulated roof deck in response to normal thermal cycles.

The last point is particularly important when precast concrete or curtain wall panels extend above the roof to form a parapet. Prudence suggests an expansion joint at the parapet – roof intersection to attenuate the differential movement.

Detail 6.9 shows a *low parapet wall detail* for use where the parapet and roof deck structures are fully integrated; it does not require an expansion joint. In this case, the wall is sheathed with plywood and capped with a metal coping cap. Aluminum pigmented asphalt paint is used to cover the exposed flashing unless a granule surfaced polymer-modified asphalt sheet is used as an exposed flashing sheet.

A scupper can be used as an emergency drain where the parapet and roof structures are meshed to act as one. Avoid scuppers in any parapet where the roof structure is likely to move independently of the parapet; if used the scupper flashing will soon be torn at or near the intersection between the systems.

I favor the use of gravel stops – or metal edge flashing – wherever the building code permits. The slightly elevated perimeter keeps the metal-to-roofing joint up out of the water. At the same time, the low elevation of the gravel stop can serve as an emergency relief for dammed storm water due to clogged drains.

While it is not shown in Detail 6.10, designing the metal deck to hang over the wall can be a big benefit on many commercial jobs where the roofing gets applied before the walls are in place. If conditions require the wall construction to be delayed, install the perimeter wood blocking with a wood fascia board to cover the blocking, the end of the deck, and blocking to wall joint. Then the roofing and flashing can be installed without waiting for the wall construction to be finished. A detail must be added to seal or provide an expansion joint between the top of the wall and the underside of the blocking.

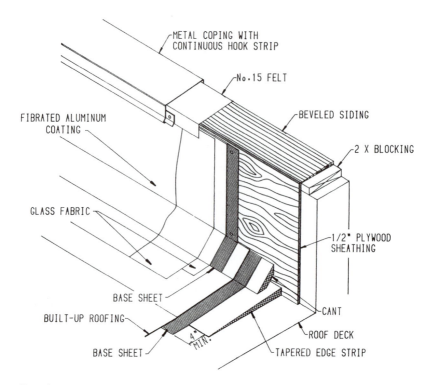

Detail 6.9 Low parapet wall.

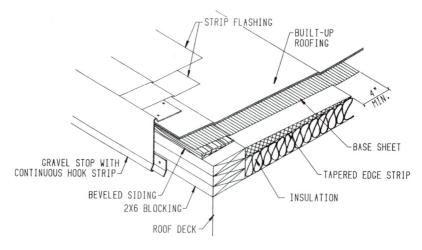

Detail 6.10 Gravel stop.

In any metal flashing, consideration must be made for the relatively high thermal expansion and contraction of the metal. Generally the length of any metal flashing element should not exceed 3 m (10 ft). Longer runs must include an expansion joint in the metal, such as shown in Detail 6.11 – a detail for a *transverse joint in the metal edge flashing*.

The details for *flashing an expansion joint* (Detail 6.12) and *flashing a relief joint* (Detail 6.13) are identical above the surface of the deck. The expansion joint detail covers and waterproofs structural expansion joints. Relief joints are installed in a roofing system to define drainage areas, to ease re-entrant corners, or other areas of stress concentration. You probably are familiar with cracks in plaster or plaster board that extend diagonally upward from a door or window opening. The corner of the window or door opening is a re-entrant corner – and an area of stress concentration. You can test this yourself. Cut two strips of paper; one strip twice the width of the other. Cut a notch half way through the wide strip. Grasp the ends of each strip with your hands and pull each strip apart. Note how much easier it is to pull apart the notched strip compared to the strip without a notch – even though the strips started out the same width at the notch. This loss of resistance is due to stress concentration.

Structural expansion joints are usually required where the direction of the structural deck changes, where the nature of the deck material changes (i.e. from concrete to steel), and where the nature of the structural support changes (i.e. roofs to walls, or new construction to existing construction). Roofing and waterproofing systems are not strong enough

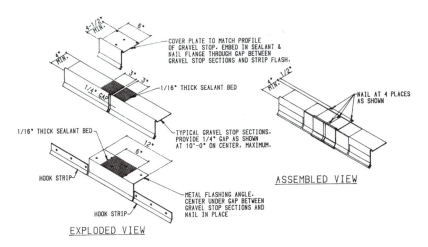

Detail 6.11 Gravel stop tranverse joint.

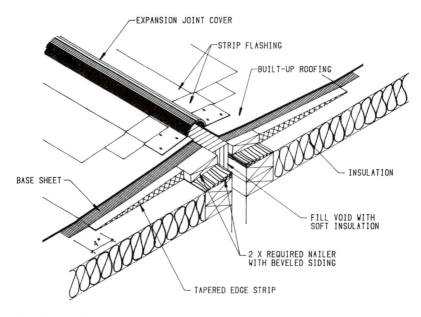

Detail 6.12 Expansion joint.

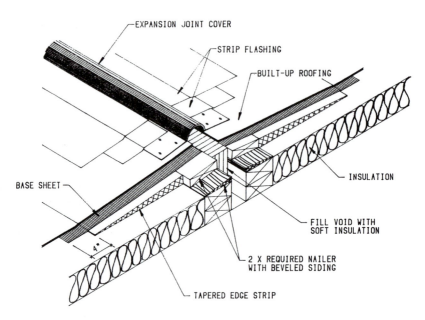

Detail 6.13 Relief joint.

to hold the building together – nor can they prevent structural movement. Remember:

> *All buildings, you know,*
> *Move like a belly dancer,*
> *But only slowly!*

Detail 6.14 depicts some of the *fastener patterns* recommended for the flanges of expansion or relief joint covers, pipe sleeves, gravel stops, edge flashing, and overflow drains. Added details show the fastener pattern for edge or gravel stop hook strips and a deck repair strip. Two rows of fasteners are used in almost every case to keep the metal flanges flat. The fasteners are closely spaced so the dimpled plane takes up some of the thermal expansion and contraction.

Coping caps such as mortar filled tile, cut stone, or precast concrete are a constant source of leakage unless there is a through wall flashing to direct the leakage outside the wall. Knee, decorative, and even structural walls are often streaked with the white stains of efflorescence from water leaching through the masonry, because the designer did not think flashing was necessary. The metal counterflashing on tall parapets must be connected to

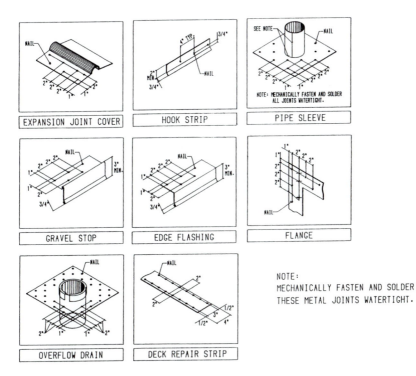

Detail 6.14 Typical nailing patterns.

through wall flashing that must extend beyond to slightly turn down the outer face of the masonry. Yes, I know the 12 mm (½ in.) of metal may be visible. If it is up high enough it will not be seen. If it is lower – live with the appearance – to let the wall stay water resistant. Another layer of through wall flashing is required directly under the coping, because the mortar between the tiles, cut stones, or precast concrete panels are not watertight. As an alternative, consider a *metal coping cap* such as detailed in Detail 6.15. Please do not rely on sealants, mortar, or lead to keep joints watertight; they just do not provide permanent protection.

Probably the worst coping cap detail is where metal counterflashing extends over half the top of the wall to end in a reglet. This guarantees water intrusion through the reglet and the unprotected surface.

Detail 6.16 is not really a flashing detail. It is included here to show how planning is needed when you attempt to combine insulation panels of differing sizes to achieve a layout so that none of the joints in the two layers match. In this case, 0.9×1.2 m (3×4 ft) glass fiber insulation panels are used with 0.6×1.2 m (2×4 ft) fiberboard panels. Without proper planning the roofing system is bound to have the through joints you are trying to avoid in the insulation system. Do not just say: one layer of "x" insulation and a top layer of "y". The size of the insulation panels and the plan for their proper installation is mandatory.

Detail 6.17 reviews how to seal off the top of metal flashing to the duct, conduit, and pipe penetrations using a bond breaker tape and EPDM tape adhered over the joint.

Obviously there are many more good flashing details than this document can hold. Your attention is directed to:

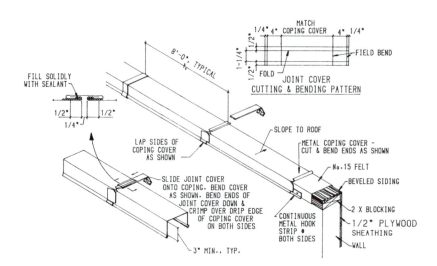

Detail 6.15 Metal coping.

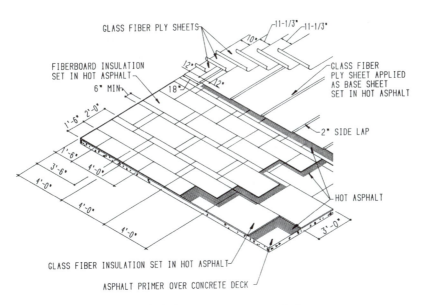

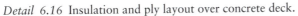

Detail 6.16 Insulation and ply layout over concrete deck.

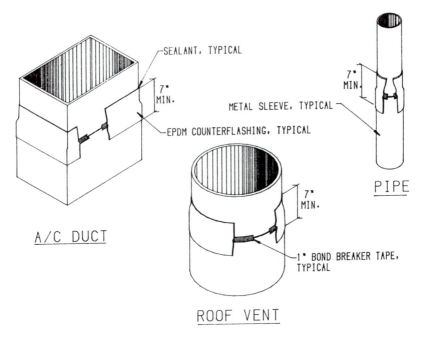

Detail 6.17 Typical EPDM counterflashing detail.

- ASTM International – Standard Details for Adhered Sheet Waterproofing
- ASTM International – Standard Guide for Design of Standard Flashing Details for EPDM Roof Membranes.

QUESTIONS

1   Flashing is the weather seal between the roofing membrane and penetrations. [a] true [b] false.
2   There is no reason to redetail the job when switching from a built-up roof to a single ply roofing system. [a] true [b] false.
3   Flashing system intersections are too difficult to draw; use the easier to draw plain sections. [a] true [b] false.
4   The roofing will tear between the end of an expansion joint cover and the nearest joint in the edge flashing, unless the two are connected. [a] true [b] false.
5   Pitch pockets should be used for [a] pipe penetrations [b] sanitary vent penetrations [c] sign supports [d] all of the previously listed [e] none of the previously listed applications.
6   Pitch pockets that are dry and incompletely filled [a] need to be filled [b] probably leak [c] both need to be filled and leak.
7   Face reglets direct water inside the flashing detail. [a] true [b] false.
8   Reglets are the easiest way to flash up against rough walls. [a] true [b] false.
9   Flashing design parameters are the same for all waterproofing and roofing systems. [a] true [b] false.
10   The easiest way to flash a curb with a single ply system is to run the sheet up and over the curb. [a] true [b] false.
11   Peer design review is inexpensive, and may save a lot of money! [a] true [b] false.
12   The US Federal government requires generic specifications to promote competition. [a] true [b] false.
13   All drains should be sumped to [a] promote drainage [b] permit the ice at the drain to melt before ice on the general roofing surface [c] Both [a] and [b].
14   All sanitary vent pipes should have a screen to exclude bees. [a] true [b] false.
15   Generally, install as much equipment on the roof as possible; this eases maintenance and reduces cost. [a] true [b] false.
16   Support HVAC equipment on wood blocking, resting directly on the surface of the roof. [a] true [b] false.
17   Prefabricated man hatch flashing should be protected from the kicking to which it is likely to be subjected. [a] true [b] false.
18   Canadians do not use counterflashing. [a] true [b] false.
19   Parapets protect adjacent buildings from wind blown gravel. [a] true [b] false.

20 Parapets tend to move differently than the insulated roof deck. [a] true [b] false.

21 Use an expansion joint flashing detail between the roof and the parapet wherever precast concrete walls are extended above the roof to form the parapet. [a] true [b] false.

22 A gravel stop detail can serve as an emergency drainage path if the interior drains or the storm sewers get plugged. [a] true [b] false.

23 Expansion joint covers are required [a] whenever the deck changes material [b] to connect new to old work [c] to relieve major re-entrant corners in the roofing [d] for any of [a], [b], or [c].

24 Fasteners for metal flashing should be in a single row, so the metal can bend like a hinge. [a] true [b] false.

25 Through wall flashing must extend out and down the exterior face of the wall. [a] true [b] false.

7 What is failure?

It seems appropriate to discuss the meaning of "failure," since this book is about roofing failures. There are a host of definitions – starting perhaps with the definition in Webster's New Collegiate Dictionary – (in part) 1. a falling short; a deficiency or lack; 2. omission to perform; 3. want of success; 4. deterioration; decay; 5. becoming insolvent; 6. a person or thing that has failed. This is surely broad enough, yet it does not seem to meet our needs except in a very general way. Here is another definition of failure:

> The AEPIC (Architecture and Engineering Performance Information Center) defines failure as: "An unacceptable difference between expected and observed performance. Performance being the fulfillment of a claim, promise, request, need, or expectation."

"Roofing failure" has been defined as:

- When a roof has outlived its designed useful life, through normal wear and tear of the elements;
- When a roof has exhibited unwanted, unacceptable, or unexpected behavior such as catastrophic leaking, splitting, blistering, sliding, blow-off, decomposition, etc. to the extent that repair costs exceed 33 percent of replacement costs, or a competent roofer cannot stop leakage through the roof.
- When a business decision is made that the roof failure is imminent and could be catastrophic, therefore immediate replacement is prudent.

Some other "failure" related definitions are:

- Durability: the capability of maintaining serviceability of a product, component, assembly and construction over a specified time. Serviceability being the capability to perform the functions for which they were designed or constructed (ASTM E632).

- Service life: the period of time after installation during which all properties exceed minimum acceptable values when routinely maintained (ASTM E632).
- Defect: the non-satisfaction of a specific job requirement (CIB – W86).
- Error of omission: a faulty human act or Act of God leading to an undesired event (CIB – W86).
- Defect: a state in which functional requirements are not met: a limit state is reached (CIB – W86).
- Damage or failure: a material disorder as a consequence of a defect (CIB – W86).
- Loss: the missing of building parts, goods, means, or individuals resulting from the damage/failure (directly or indirectly) (CIB – W86).
- Costs: the financial consequences of a loss (CIB – W86).

Some examples of acknowledged roofing failures include:

- The management of a psychiatric hospital felt the roof failed because some of the patients were trying to cut one another with pieces of fake slates that fell off the roof.
- A senior citizen felt her roof failed when the wood fiber-cement shakes turned to mush in about five years of the 50-year manufacturer's warranty.
- A school superintendent felt the roof failed when leakage first appeared before the roof was finished. Leaks continued all over the school, requiring another shift of maintenance personnel to move the buckets every time it rained.
- The president of a company had the roof replaced when water leaked into his private lavatory – the leak fell directly into the commode; it never even got the floor wet.

Frequently failures are made evident by consistent leakage, usually started when the building is still under construction, and continuing despite the best efforts of skilled roofers.

Perhaps there are as many definitions of failure as there are failures. I prefer this definition: Failure occurs when the owner does not get the service or service life for which he bargained.

Failures are due to
Getting less than expected
From the agreement.

However difficult to define failure, it seems to be easily recognized. I have conducted several surveys on roofing failure. I was always careful not to define failure, but still received answers from all over the United States and Canada without getting one request for a definition.

Most of our roofing systems are so trouble free that it frequently takes the work of many people to cause the system to fail. With multiple contributions to failure, who's at fault? With several people or groups of people attempting to contribute to a failure, I find the time line of the work a useful tool in attempting to assign responsibility. Identifying when a failure occurred may sometimes be important and more difficult than defining the failure.

Many investigators rely on "laundry lists" of defects performed by each party, regardless of when the error occurred. Unfortunately, they fail to link the items in the laundry list with the actual failure.

> *If you examine*
> *Anyone's work close enough*
> *You will find errors.*

Lawyers tend to like the investigators addicted to "laundry lists" of errors because they tend to involve everyone on the job as defendants; even if there is no cause and effect link between the itemized error and the failure. I can't tell how often a designer has been blamed for splitting asphalt-glass fiber shingles because he did not provide "adequate ventilation" below the roof deck.

> *Asphalt-glass shingles*
> *Often split when they are not*
> *Properly designed.*

Or, the many roofing contractors blamed for the wind loss of seal-tab shingles due to "improper nailing."

> *On seal-tab shingles*
> *The nail location means little*
> *If the sealant fails.*

The investigator can often make more supportable judgments by placing the departures from normal practice on a time line. Consider a roof so poorly designed that it would have failed even if it had been installed by angels. Since the design work preceded the installation, any but the most egregious errors on the part of the contractor are moot.

Think of the manufacturer who blames the contractor's installation and homeowner's maintenance for the failure of a product that lasts only a few years, but is warranted for 25 or 50 years of service. Here the designer has some responsibility (the design happened first), but the designer probably relied on the special expertise of the manufacturer. The reliance was misplaced. Nothing the contractor did has a significant influence on the service life of the product, and no degree of maintenance would enable the product to survive for the warranted term. There's a good argument here that the product failed when the manufacturer thought of it, because no subsequent actions were going to significantly influence the product's performance.

Roofing failures are usually due to:

- inappropriate constraints and other actions by the owner,
- design errors,
- defective construction work, and
- inappropriate materials.

The *owner* becomes part of the problem when he provides inadequate funding, selects design professionals for reasons other than their competence, insists on using new, fashionable materials (read: untried), sets himself up as the principal technical authority, contracts with the lowest bidder contractor, and refuses to provide for any quality control or oversight.

For many jobs, such as a simple warehouse, or re-roofing a small simple building, a design professional is hardly necessary. But, in that case, prudence suggests relying on a good roofing contractor's detailed proposal, and having his proposal reviewed by people competent in roofing technology. It probably would be beneficial to have the roofing technologist present at the pre-job meeting, available on call (in case of trouble), and available to check the final punch list and close the job down.

The error most purchasing agents usually make is hiring a new roofing contractor who lacks the experience, equipment, and personnel to perform the work in a reasonable interval. The scope of work is usually generic, lacking in detail and does not include a date when the work will be started and concluded.

Some owners, like the US government, require generic specifications. For example, a designer cannot write "stainless steel" flashing on the detail drawings. "Metal" must be shown on the drawings, and "stainless, galvanized steel, aluminum, copper, or equal" might appear in the specifications. The stated goal is to stimulate competition. What is accomplished is to remove the responsibility for the choice of the metal from the designer. The government might respond that the designer must approve or disapprove the choice submitted by the contractor, but I have found that the material choice is often made by the government's representative – not the designer.

Some of the typical Federal specifications require the contractor to hire an inspector for the work. I support the concept of a competent inspector to monitor every major roofing job, but he should not be hired by or paid by the roofing contractor whose work he is monitoring. Do you really want the fox in the hen house?

Roofing monitors
Must be independent of
Roofing contractors.

Generally *defective designs* are the largest source of roofing failures. Most architects have a wonderful vision of the space they wish to create, but have little or no knowledge about how to keep the space watertight. Engineers know all about stress–strain, but often neglect to slope the roof to drains and forget that:

> *Holes and gravity*
> *Are the primary sources*
> *Of all roofing leaks.*

Often the least experienced designers are charged with the task of selecting the roofing system and developing the details. Almost none of them have received any training in roofing or materials technology, and must rely on the latest or most persuasive salesman or the Internet for guidance. Responsible sales personnel are worth their weight in gold, because they can be a source of intelligence and knowledge, but as in any endeavor, only few have been employed long enough to have the experience and who have developed the judgment needed to maximize their value.

> *Good roofing systems*
> *Depend on how they are piled.*
> *Not the material.*

Ignore the product or systems that can be installed by anyone, that can be installed over snow or water in any form, that will last forever and that are in the fore-front of technology. You are urged to see every salesman who calls on you. Never be too busy. Remember:

> *Curiosity*
> *Is a proven path to truth.*
> *Ask lots of questions.*

Contractors are a resource to be valued and cultivated. They are aware of the availability of materials, and often are the first to be aware of materials or systems that are giving problems in the field. Roofing contractors are often small organizations, although some consolidation of smaller firms has been experienced in recent years. They are generally politically conservative and slow to change. As always, a few contractors are outstanding. In general, the contractors in the mid-western and far western states are the most professional, followed by the contractors in New England. Contractors in the deep-south are the lowest on the professional ladder of competence. Interestingly, the direct labor cost unit of construction is approximately constant over the nation. The higher productivity roofers are earning more than their lower productivity contractors, so the labor cost per square metre or per roofing square is approximately the same in New Jersey or Mississippi.

Where defective design work contributes more than 50 percent of the roofing failures, defective workmanship accounts for about 30 percent of the total roofing failures. Most of these errors by contractors are due to ignorance about the consequences of their actions and absence or poor supervision.

As an example: The owner asked me to investigate the condition of a newly installed built-up roof that replaced an older roof on a shopping center. The owner was suspicious because a crew of roofers appeared on a holiday weekend (after he had been after the roofer for over a month to complete the job) and they completed their work before the weekend was over. Test cuts revealed that the steel deck was devoid of adhesive or insulation fasteners, and the roofing membrane contained two of the four felt plies the contract required. The owner of this contracting firm was a good, honorable contractor. The foreman and superintendent were selling the materials that they were not installing to others without the owner's knowledge. Needless to say, new control procedures were established while the foreman and superintendent sought new positions elsewhere.

Roofing is physically hard work that takes place in areas furthest removed from creature comforts; where the weather is too cold, too hot, and sometimes too wet. Each crew hopefully has at least a few skilled roofing workmen; these usually are too busy to supervise others on the roof. Skilled personnel are usually supplemented with less experienced transient labor. Daily prisoner release programs in some states supply the additional manpower. Combining inexperienced workers, with inadequate supervision, with the emphasis on production rather than quality almost guarantees problems. Good effective roofers have both high quality of installation with a high rate of production.

Materialmen or manufacturers contribute their share to the roofing failures by:

- overselling the merits of their products or systems,
- ignoring fundamental product or system weaknesses,
- failing to provide appropriate recommendations,
- failing to enforce their recommendations, and
- emphasizing new products, rather than products or systems of proven worth.

The warranty offered by many manufacturers is often much longer than the length of time the product has been in existence. When questioned about this practice, the response often is that the warranty is like the price; it is due to competition; it is negotiable – it doesn't represent the service life. This sounds reasonable, except the average consumer believes the warranted service life is the minimum service life the purchaser can expect. Indeed, the typical consumer believes the product should last well past – perhaps 1.5 times the warranted service life.

An interesting question is based on the fact that there is no test or batch of tests that can accurately predict roofing system performance. Since the manufacturer can't predict performance, and the consumer is enticed to buy based on the warranty, don't the manufacturer's lengthy warranties constitute fraud? I would be very happy to see all warranties disappear.

Every so often a new product or system is proposed that contains an inherent flaw that is ignored by the marketers. Here are two examples.

In the 1950s, a respected manufacturer of building materials promoted a new system. This new system used two plies of asphalt-coated organic felt instead of the four plies of asphalt organic felt then in use for most built-up roofs. This new system was promoted by the manufacturer as: "1 + 1 = 4." Other manufacturers jumped on the bandwagon after they assumed the originator had done the fundamental product development. The originator relied on tensile test results that showed that two heavier weight plies were just as strong as four thinner plies. This reliance was misplaced. Load strain tests of the completed membranes showed the membranes with the fewer thicker felts were significantly weaker than the conventional membranes. The result was a disaster for the roofing industry as the roofs started to fail.

Part of the blame for the failure was due to the new product-promotional activities that are typical in the roofing industry, including:

- heavy advertising investment,
- promotion to marginal roofing contractors who often have a low capitalization, experience, and trained manpower (established contractors are too conservative and have seen too many new systems fail),
- promotion to price sensitive customers such as shopping centers, and
- promotion of phased construction technique (installing one ply over the entire area before installing the second ply).

The phased construction technique made it simple to dry in the job (provide temporary protection), but the exposed felts absorbed water that the subsequent ply sealed into the membrane – resulting in massive blistering.

A second example of marketing myopia, started with the identification of the problems associated with asbestos. This led to the collapse of the asbestos-cement industry in the United States. Some asbestos-cement product manufacturers substituted wood fibers for the asbestos-cement, ignoring the capacity of wood to absorb water, swell, lose strength in an alkaline environment, embrittle though carbonation from the cement, and rot. Each manufacturer has tried to use wood fibers in somewhat different ways to make fake slates or shakes, and each product has failed in its own way. Some slates curl upward (called "cupping" or "potato chipping"), some turn to a mush, and micro-cracks in others broadcast the way through the slate. The durability of these products depends very much on the durability of the protective coating provided. Some fake slates have an acrylic glaze coating to make them shiny, others have pigmented paint coatings on one or both sides, and the surviving product left on the market has a thick, hard coating resembling baked enamel.

In hand, many of these wood fiber-cement products are quite attractive, are lighter in weight than tile or slate, and offer better fire resistance than wood shakes. This good in hand appearance, plus warranties ranging from 25 to 50 years, persuaded many people to purchase these once in a lifetime products. Dissatisfied homeowners may get these defective shakes repainted at the manufacturer's cost, but repainting is a mere band-aid that does not address the fundamental problem of high water absorption. One of these shakes absorbed a remarkable 148 percent of its dry mass; perhaps enough to collapse some roofs.

The failures of so many of these fake shingles have blighted the market for other new products of plastic and rubber, aimed at the slate and tile market. These new products are very attractive; they generally do not have a Class A fire rating, and do not have a long exposure history. Some appear to be worthy substitutes for natural slate. I recommend a high quality laminated asphalt-glass shingle if Class A fire resistance is required.

> *Materials don't fail*
> *They obey physical laws*
> *We don't use them right.*

In some cases, manufacturers provide little or no information about their product or how it should be installed. Worse, some neglect to list the product limitations – the latter is characteristic of many products made outside the United States. Neglecting or hiding product limitations is not the way to increase the long-term sales or market penetration, because any manufacturer's sales are mainly the result of his long-term relationship with his customers. I'm happy to say that this lack of product information has changed – particularly with the advent of the Internet. Perhaps we now get too many words and not enough information.

> *Used like a lamp post*
> *For support instead of for*
> *Illumination.*

Occasionally, the manufacturer's instructions are not followed. When this is pointed out, the local salesman sometimes advises: "That requirement isn't necessary – and, what are you worried about? – we're still sending you the warranty!" Be very careful. The specific requirement may be unnecessary, or the salesman may be protecting his customer to the detriment of the job.

Innovation has been both a blessing and a source of failure. In many companies management is required to come up with something new about the time the new Five Year Plan or the budget is assembled – to stave off the criticism about the new items in past plans that didn't work. "New" is required to obtain the funding needed for marketing. "Old" is unexciting and either doesn't need funding, or surely does not require the special funding needed for a new product or system introduction. "New" means you get to keep your job. "Old" means a lack of imagination, of progress (progress comes only through change) and, of effective management. In building materials, unlike computers, none of these concepts should be operational.

In computers, the current market seems to be willing to believe it's obsolete if it works. A new computer is considered old if it is only three-years old. A three- or four-years old printer is considered obsolete and no longer has drives available for the newer operating systems.

Building materials are expected to perform, and for a very long time. Short-term performance is not acceptable, and is remembered for a very long time. Manufacturers who promptly correct their mistakes are fondly regarded. Manufacturers who try to bury their mistakes promptly lose respect and market position.

Having spent many years in research, I'm surely not against innovation. I just don't believe in irresponsible innovation that uses the public as uninformed experimental subjects, or continued "cost improvements" until the product or system doesn't work.

New building products must have extensive pre-introduction evaluation, and at least five years of blameless service as experimental applications where they will be used. I am aware of the intense pressure to quickly commercialize a product, but my experience teaches, that despite our best efforts, most experiments fail.

Always remember:
Mother Nature is a bitch
She attacks the weak.

Once, while I was employed by a material supplier, I was asked by the vice-president in charge of manufacturing about the status of the research on a new product. I told him that we completed our laboratory work; we were awaiting the results of our outdoor exposures. "Put another man on it" was what he told me. Failure-free roofing is possible when all parties involved freely and openly communicate, and help each other toward that goal.

QUESTIONS

1   Failure occurs when the purchaser does not get the service or service life for which he bargained. [a] true [b] false.
2   Most roofing failures are the result of a combination of poor design, inadequate workmanship, and inappropriate materials. [a] true [b] false.
3   Less than 50 percent of the roofing failures are due to poor design. [a] true [b] false.

4 [a] People [b] Materials [c] Both people and materials cause roofing system failures.

5 Laundry lists of defects point out the cause of failure. [a] true [b] false.

6 Improper maintenance seldom is a sole cause of failure. [a] true [b] false.

7 The designer has the responsibility to select or approve the specific materials and systems used on the job. [a] true [b] false.

8 The time line of the work is sometimes useful in establishing fault responsibility. [a] true [b] false.

9 Professional designers are necessary for every roofing job. [a] true [b] false.

10 The advise of salesmen and contractors should be ignored. [a] true [b] false.

11 Any good roofing system can be installed in any weather. [a] true [b] false.

12 An emphasis on production rather than quality causes some failures. [a] true [b] false.

13 Good roofers emphasize quality; production doesn't mean anything. [a] true [b] false.

14 Some failures are caused by overselling features of a product. [a] true [b] false.

15 Product weaknesses should be hidden to maximize sales. [a] true [b] false.

16 Some products have an inherent flaw that makes them failure prone. [a] true [b] false.

17 Innovation tends to increase failures. [a] true [b] false.

18 All new products or systems must prematurely fail. [a] true [b] false.

19 Both laboratory testing and field trials are needed before product introduction. [a] true [b] false.

20 Failure free roofing is possible. [a] true [b] false.

8 Performance vs prescriptive specifications

There have been many design fads over the years. I must admit that the highest emotional levels for these fads seem to be found among the students, young designers, academic personnel, and national laboratory personnel; older designers may pay lip service to these fads to satisfy clients, but otherwise intelligently ignore them. Currently, the following buzz words are active and are taken quite seriously and emotionally by a lot of designers:

- Performance specifications,
- Sustainable construction,
- Green construction,
- Global warming, and "Cool Roofing."

Many individuals seem to feel we can correct all of our problems if we only had good *performance specifications* instead of the old fashioned prescriptive specifications most of us are currently using. Performance specifications list the outcome desired. Prescriptive specifications list the work and material needed to gain the outcome desired. The question is, how are we most likely to achieve the end result we desire?

We have quite a few performance specifications in the roofing field including:

- ASTM International – D3018 – Class A fire resistant shingles.
- ASTM International – D3161 – Wind resistance of shingles.
- ASTM International – D3746 – Impact resistance of roofing systems.
- ASTM International – D4073 – Tensile-tear resistance of roofing systems.
- ASTM International – D4932 – Fastener rupture and tear resistance of roofing.
- ASTM International – D5385 – Hydrostatic rupture resistance of waterproofing.
- ASTM International – D5601 – Tearing resistance of roofing and waterproofing.
- ASTM International – D5602 – Static puncture resistance.
- ASTM International – D5635 – Dynamic puncture resistance.

I suspect few of these test methods or standards are used by any roofing system specifiers, even if they are very much in favor of performance specifications. Most specifiers rely on the prescriptive specifications recommended by one or more of the manufacturers.

The performance specification proponents seem to desire a simple specification that covers all contingencies. Let's assume that there are specific performance tests that can be relied upon. If we specify these, they become prescriptive standards, so the whole discussion of performance vs prescriptive standards becomes bogus.

In reality, there is no test or group of tests that predicts roofing system performance, so we must stick with our existing imperfect system, and continue to develop standards and criteria that tend to maximize performance.

A performance spec.
Needs tests to show the future
A cloudy glass ball.

The concept of *sustainable construction* is more recent than performance specifications. While not clearly defined, it generally means to design using materials that can be reused after the service life of the original design has been satisfied. This concept has some merit if you are going to select one of two materials that have equal performance and one can be reused and the other discarded.

We currently have two problems. These are our current experience with recycling debris, and our ability to accurately foresee the future.

Many towns recycle their waste in an effort to reduce material landfilled. Waste paper dominates the volume recycled; a small quantity – mostly newsprint – is recycled into cardboard, fiberboard, and similar products, but the largest volume is used for fuel to generate electricity. The economics are not very attractive, and recycling has been halted in some areas, because it costs more to recycle than to use virgin materials or regular fuels. Recycling for fuel will not be attractive until the cost of conventional fuels is significantly increased. Asphalt roofing could be a good fuel, if the cost of conventional fuel gets high enough, and if controlling authorities permit fossil fuel combustion. Plans are being made now to recycle thermoplastics such as PVC and TPO, but there are many technological problems to overcome before this becomes economically feasible.

The other problem with sustainability is that our roofing systems are designed for long life; we have no knowledge about the market and technological needs even ten years from now, much less twenty or more years from now. It would seem that our efforts should be to provide the best roofing system possible; the one with the longest life.

Green construction is another new buzz phrase; it is related to sustainability. If we are going to have a garden on our roof, our structural capacity will have to be increased to handle the increased dead load, our roofing system must be replaced with a waterproofing system, a drainage system, a grass cutting or weeding system, and a watering system. Leaks that are currently an inconvenience will become much more serious, be more difficult and costly to reach, and are likely to cause more subsequent damage than would occur with our old roofing system.

Most of the "green construction" advocates really do not expect us to have garden roofs. They are more for design strategies that minimize the use of energy by careful site selection, water efficiency, energy efficiency, and materials conservation. These are good objectives, but again they are very difficult to achieve given today's low cost of fossil fuels.

Somewhat related to the theme of green construction is the *cool roofing* idea from one of our nuclear laboratories that suggests we should paint or color our buildings white to reduce the heat island effect in cities. The result their calculations suggests is a significant reduction in cooling energy requirements, because our cities would not absorb and reradiate heat, reducing our peak air conditioning cooling costs, and fossil fuel consumption. Cool roofing may slightly reduce the fossil fuels consumed, but since capital cost is the major cost of electricity, and conservation provides fewer units of power to cover the capital cost, the price of electricity must rise with the conservation effort. Again, the best long-term solution is the significant increase in the cost of fuel, because the elastic demand will reduce the usage.

I have long been an advocate for using the local dirt color on roofs. Dirt colored roofs blend into the landscape, and don't show any dirt. The cost of painting things white every few years would more than make up any savings from decreased cooling costs. Unmentioned is the cost of medical treatment for individuals blinded by the reflection from a white roof that they overlook. We would probably have to perform studies to be sure adjoining roofs were not focusing their reflected energy where they could do harm.

One state was quite convinced about the reduction in energy needs from coloring the roof white. They changed the building code so that you could use less insulation if you had a white roof. Like most government plans – reducing the insulation was not the planned outcome.

Cool roofing tends to favor manufacturers who make white roofing. This may be inappropriate because many black systems will outlast white systems. For example, I suspect black TPO will outperform white TPO, just as black EPDM outperforms white EPDM.

The electrical energy producers support this white plan because they don't have to invest in additional capital equipment if their peak loads are moderated. Of course, they could moderate their peak load by changing their multiple rates to a single rate for all clients. The maximum usage clients would find ways to reduce the peak load. Additional savings can be achieved by firing the managers, accountants, and billing clerks – to say nothing of the taxes saved when the rate regulator departments are fired.

There is little credible evidence that *Global warming* exists, outside of normal cyclic variations, and the Kyoto agreement recently rejected by the

United States offered no significant solution to the green house gas problem. Caring for our atmosphere on this singular planet is important, and we should take whatever steps necessary to assure constant improvement. We should also limit burning of fossil fuels because they are a treasury of chemical raw materials too valuable to burn. Raise the price of fossil fuels. That will reduce usage, increase the search for alternative energy, and foster improved energy efficiency. Throughout our history, our greatest advances were made as we discovered and harnessed new sources of energy – why can't it continue?

I feel it is boastful to blame global warming (if it exists) on human activity. Ambrose Bierce once defined the world as: "A sphere of matter. Two thirds of it is covered with water. It was made for Man; who doesn't have any gills."

QUESTIONS

1 Currently, there are no performance specifications for roofing. [a] true [b] false.
2 If effective performance specifications exist, they will be prescriptive. [a] true [b] false.
3 Sustainable construction means building temporary reusable buildings. [a] true [b] false.
4 Great economic savings currently result from recycling. [a] true [b] false.
5 We do not know what materials will be required in the future. [a] true [b] false.
6 Growing grass on the roof will make it last longer. [a] true [b] maybe [c] false.
7 Green roofing is less expensive than conventional roofing. [a] true [b] false.
8 Cool roofing favors those manufacturers who make white colored roofing. [a] true [b] false.
9 We should limit the burning of fossil fuels because they are too valuable. [a] true [b] false.
10 Global warming is not probably the result of actions by humanity. [a] true [b] false.

Part II
Case studies

INTRODUCTION

As Peter Green reported in the February 1987 of Architectural Record: "Roofing may be one of the least rewarding aspects of an Architects design program, but it deserves undivided attention or else it can become the most conspicuous of the program's shortcomings. (Unfortunately, buildings are often remembered in the industry, not for their design awards, but for their roofing litigation.)"

Water has long been our enemy. Alec Tiranti, in his *Ten Books on Architecture* quoted Leon Basta Alberti (1695):

> *For Rain is always prepared to do Mischief and wherever there is the least crack, never fails to get in and do some hurt or other. By its Subtlety it penetrates and makes its way, by its Humidity rot and destroys, by its Continuances loosens and unknits all the Nerves of the Building and in the End ruins and lays waste.*

Each of the next short chapters contains one or more actual case studies in roofing technology that illustrate some type of general roofing failure. Without presenting the actual location or the real names of the parties involved, you will be given all the job information, field observations, and laboratory test data. In some cases, you will be given theories about the cause and responsibility for the failure, advanced by interested parties. You will then be asked some pertinent questions intended to be answered in class room discussions and, if desired, essays.

The case in Chapter 9 is presented in more detail than the balance of the cases; the others rely on summaries of the facts and positions.

The answers section at the rear of the book provides general information about the real world outcome of each case where it is known to me. Be advised; I don't always agree with the outcome in every case – sometimes

Blindfolded justice
The statue on the court house
Is unseen in court

9 The case of the leaking book warehouse

Four years before our investigation, a major book publisher had a distribution center constructed in the hinterland where it planned to take advantage of the lower labor rates to warehouse and distribute the many books it produced. Unfortunately, the roof started leaking during the first spring following the first winter after the building was occupied. Various patching techniques temporarily halted some of the leakage, but the leaks almost always returned.

The contract drawings show that the warehouse forms a rectangle in a 60×120 m (~200×400 ft) plan. Steel frame and cantilevered bar joists support a steel deck. The building is quartered by CMU fire walls and masonry in-fills the perimeter walls. The perimeter walls are 10 m (~33 ft) tall, forming a 1 m (~3 ft) high parapet.

The contract specifications require:

- a plastic vapor retarder adhered to the steel deck with a propriety fire resistant adhesive,
- a 38 mm (1½ in.) layer of perlite roof insulation, adhered to the vapor retarder with the same fire resistive adhesive,
- a top 38 mm (1½ in.) layer of perlite roof insulation adhered to the first layer with hot asphalt, and
- a four ply asphalt-organic felt gravel surfaced asphalt built-up roof.

Roof leaks begin shortly after any rain storm; they are located predominantly, but not exclusively, near the perimeter and fire walls.

GENERAL OBSERVATIONS

The 7200 m² (~778 square) roof area is quartered by firewalls. The roofing is a gravel surfaced asphalt built-up roof that has no visible slope. Each roof area is served by two 150 mm (6 in.) diameter drains. The roofing terminates below metal counterflashed asphalt built-up base flashing on the parapets and fire walls. The base flashing is displaced away from the walls; it has shear (diagonal) folds pointing toward the center of each roof segment. In some areas, there are four linear marks, about 6 mm (~¼ in.) on the face of the base flashing, parallel and close to the lower edge of the counter-flashing. The top of the base flashing is pulled from beneath the counter-flashing in some areas. The base flashing has tears perpendicular to the shear folds. There are a few linear repair patches on the roofing membrane.

OBSERVATIONS DURING SAMPLING

We selected Sample A at random (near no obvious membrane defects), near the northwest corner of the northwest roof section. The top-to-bottom roof construction is:

- A gravel surfaced asphalt built-up roof adhered to two 38 mm (1½ in.) thick layers of perlite insulation adhered together with hot asphalt. The underside of the bottom layer is adhesive free. These components feel dry.
- A red plastic vapor retarder sheet with pencil width (showing that they had never been compressed) lines of adhesive, about 150 mm (6 in.) apart, parallel to the length of the corrugated steel deck.
- A conventional 38×150 mm (1½×6 in.) corrugated steel deck, with 50 mm (2 in.) wide ribs and a 100 mm (4 in.) wide upper flange, with a reinforcing corrugation down the center of the top flange. The deck's vertical section is "M" shaped (the top flange is not flat). Black parallel lines of adhesive, 150 mm (6 in.) apart, run perpendicular to the steel deck flange and are flattened at the points of the "M" shaped deck. The bottoms of the ribs are rusty and contain water which holds flakes of rust.

We selected Sample B near the center of the northwest roof segment. The observations are the same as recorded for Sample A, except that the steel deck is dry.

We selected Sample C near the fire wall and northern parapet. The observations are the same as recorded for Sample A.

We selected Sample D to include a patch on the roofing in the northeast quadrant. The patch is over a split in the roofing directly over a through joint in the top insulation layer. The asphalt between the insulation layers is absent. All components feel wet. The steel deck is rusty.

We selected Samples E and F in the southwest and southeast quadrants respectively, near the perimeter of the roof. The observations are identical to those recorded for Sample A.

Our laboratory data are shown in Table 9.1 and summarized as:

- The asphalt-organic membrane analyses show that most of the components are within expected parameters, except that the average interply asphalt is slightly low in Samples A, C, and D.

Table 9.1 Laboratory report

Membrane analyses	Sample designation						
	A	*B*	*C*	*D*	*E*	*F*	*Typical values*
Adhered aggregate, kg/m	11.6	9.1	12.3	–	11.0	11.8	9.3–17.2
Asphalt top coating, kg/m	2.7	2.3	3.1	–	2.6	2.9	2.2–5.5
#15 asphalt-organic felt plies	3.1	3.0	3.2	3.2	3.1	3.0	≥3
Asphalt-coated base sheet plies	1.0	1.1	1.0	1.0	1.1	1.0	≥1
Mean interply asphalt, kg/m	1.1	1.2	0.8	1.1	1.2	1.2	1.2–1.5
Mass % moisture content							
Membrane	0.9	0.8	3.4	13.8	3.1	3.2	2.9–5.6
Top perlite insulation	1.0	0.9	1.1	98	1.8	3.2	1.7–5.0
Bottom perlite insulation	8.2	1.1	1.4	110	2.0	4.1	1.7–5.0
Observations during analysis							
Condensation in sample bag	yes	no	yes	yes	yes	yes	(no)
Split membrane	no	no	no	yes	no	no	(no)
Split type	–	–	–	Λ	–	–	(–)
Interply voids							
Holidays	no	no	no	no	no	no	<3
Failure to embed	no	no	no	no	no	no	<3
Blister	no	no	no	no	no	no	<3
Pinholed asphalt	no	no	no	no	no	no	no

- The roofing membranes in Samples A and B are dry. The membrane in Sample D is wet; the rest of the membranes (Samples C, E, and F) are moist.
- The perlite insulations from Sample D are very wet. In general, the lower insulation layer contains more water than the upper layer.
- We found moisture beading on the inside of the sealed sample bags in five of the six samples, showing free water within the system.
- The membrane in Sample D had an Λ split, showing that the splitting force came from below the membrane.

The following statements were made about the failure by interested parties. The *designer* agreed the roof failed. He stated that his design met all the current standards for good roofing design. Specifically, he specified the use of a vapor retarder and adhesive required to minimize flame spread. (These materials were required to get a Class 1 fire rating from the insurance company, ever since the very severe fire at a General Motors plant at Livonia, Michigan. Investigators report that the Livonia fire was spread by the asphalt insulation adhesive melting and running through the steel deck to feed the flames.) Conventional roofing was specified. He had no duty to supervise the general contractor or the sub-contractors. The failure must be due to the work by others. He mentioned the "M" shaped deck (showing the deck was not properly installed), the incorrect installation of the vapor retarder adhesive across instead of with the top flanges in the steel deck, and allowing the insulation adhesive to dry before setting the insulation.

The *general contractor* agreed the roof failed. He pointed out that there were several reports of wind loss failures traced to the poor attachment of the insulation using the plastic vapor retarder system. In addition, the designer failed to slope the roof to the drains, and failed to require expansion joints at the fire walls and elsewhere to divide the 30×60 m ($\sim 100 \times 200$ ft) roof areas into smaller segments. In addition, the two drains in each roof area were undersized; they should have been either 200 mm (8 in.) in diameter, or there should have been more drains.

The *roofing contractor* agreed that the roof failed. He blamed the "M" shaped deck and the plastic vapor retarder system for the poor attachment. He installed the vapor retarder adhesive across the deck flutes because that is the only way he could get them attached. He also said that the steel deck specified had a reinforcing corrugation down the center of the top flange – just where the adhesive bead was to be installed – making it unsuitable for use with the plastic vapor retarder system. He pointed out that the insulation adhesive dried on top of the vapor retarder because the "M" shaped deck and reinforcing corrugation failed to support the plastic under the adhesive.

The *material man* (supplier) allowed that the roof complied with all of the material manufacturer's recommendations.

All parties agreed that the roof failed because of lack of attachment. Thermal cycles caused the system to move toward the centroid of each roof segment by ratchet action.

QUESTIONS

1   What are the cause and effect factors critical to the failure?
2   Which factors are of no or limited importance?
3   Assign a proportion of responsibility to the work of the designer, general contractor, roofing contractor, and material man.
4   How can this type of failure be avoided?

10 The case of the shattered slates

We were called in by the general contractor. The job was a new hospital facility for mentally disturbed children. As required by the specifications, a roofing contractor was installing imitation slate shingles provided by a manufacturer with a lot of experience in asbestos-cement products.

Due to the threat of asbestosis, the manufacturer had switched to a formula including cellulose instead of asbestos fibers and Portland cement. These fake slates were prepared by dry mixing cellulose fibers (recovered from newsprint), Portland cement, fumed silica and pigment, transferring them onto a moving belt, spraying the mixture with water, and pressing the matrix to form a sheet. Subsequently the sheet was autoclaved, slates were cut to size, and nail holes punched. The final finishing operation used an acid to wash efflorescence off the shingles and a final clear acrylic coating was applied to give the shingles a shiny appearance.

Flexural strength tests were an important part of the quality control program. Specimens were loaded in a three-point bending program. After the load increased to a peak and then declined (showing the specimen had broken), the peak load was recorded, and the specimen was removed from the testing machine. Curiously, there was no visible crack in the tested specimen, but when flexed even slightly, the specimen fell apart. This observation and the fact that the deflection at break was not routinely recorded is significant, as we shall subsequently appreciate.

These fake slates have a modulus of rupture of 12.5 MPa (1820 psi), a handleability index (a measure of brittleness) of 1.5, and a deflection at break of 1.9 mm (0.075 in.). These are among the strongest and most brittle of the artificial roofing slates and shakes we have tested.

As with natural slates, these artificial slates are to be installed by "hanging." This means the nail is not fully driven; the slates are allowed to hang on the shanks of the nails. The problem becomes clear when it is realized that the thickness of a typical roofing nail head is ~2.4 mm (0.094 in.) – more than the deflection at break. This means it is entirely possible for a shingler to break a shingle by standing on it and not even be aware of the break, because of the peculiar tendency for the crack to be invisible until further movement occurs due to snow load or whatever.

With carbonation due to moisture and exposure (calcium hydroxide is washed from the cement into the cores of the cellulose fiber where it is converted to calcium carbonate), the slates loose strength to a modulus of rupture of 4 MPa (582 psi), and decrease their deflection at break to 1.1 mm (0.042 in.). This makes repairs difficult, since any traffic is likely to break more slates.

Prior to the job at the hospital, there had been a large number of slate cracking complaints with this new product, not publicized in the industry. The original manufacturer was forced into bankruptcy, and the assets of the company were purchased by a major building materials manufacturer, who resumed production after a short interval.

The first complaint at the hospital, about half way through the application to the multiple buildings involved, was for blotchy appearance. The new manufacturer blamed the contractor for not following his recommendations to mix up the slates from different pallets before installation. Mixing heavy slates is quite different from mixing floor tiles from different boxes.

A great deal more emotional second set of complaints cited shards of slate that fell off the roofs into the exercise yard. Some of the patients were attempting to cut one another with the shards. This guaranteed a reaction from the locals and the state.

Our field investigation confirmed the blotchy appearance – it was far greater than would be overcome by mixing slates from different pallets. We took many pictures of the blotchy appearance and of the broken and missing slates. In addition, we obtained slates that had never been exposed to the weather, for our laboratory examination.

Our laboratory tested and examined the slates we obtained in the field and found micro cracks in the face of the slates. The micro cracks contained the acrylic top coating applied in the factory, showing that the cracks originated in the manufacturing process and not due to any subsequent handling.

These micro cracks serve as natural points of stress concentration. They are quite likely to expand and go deeper with normal thermal cycles, as has been observed with other brittle materials with surface crazing.

QUESTIONS

1   What are the cause and effect relationships within this roofing failure?
2   What is the relative contribution to the failure by the design, workmanship, or material man's efforts?
3   How can you avoid this kind of failure?

11 The case of the department store's splitting return

We were asked by a major insurance company to examine the roof on a New England department store. The following summarizes our initial notes, observations, and the results of our laboratory studies.

In the field, we found the store had roofs at two elevations. The principal roof area, over the retail store, had an area of about 4600 m² (~500 squares). The second roof area, over a storage area on a mezzanine, had an area of 280 m² (~30 squares). Both roofs were less than a year old. The building was conventionally steel framed, with the roofing supported by bar joists and a steel deck.

The bottom-to-top composition of the roofing system was:

- a 0.71 mm (22 gage) wide rib steel deck spanning 2.1 m (7 ft),
- a conventional asphalt-kraft paper vapor retarder ribbon mopped to the steel deck,
- a 76 mm (3 in.) thick layer of glass foam insulation hot mopped to the vapor retarder, and
- a four ply conventional built-up roofing membrane with a gravel surface.

The roofing membrane was split at several locations parallel to the bar joists supporting the roof. Test cuts showed that the membrane splits were directly over the through joints in the glass foam insulation, and were aligned with the length of the roofing felts. Each membrane split was near a bar joist, and the vapor retarder membrane had not been split.

All the components were wet to the touch in the vicinity of the splits, but firmly adhered to each other. We saw a few optimistic flashing details (we hope they work), but none of these were related to the splits we observed or the leakage reported.

Our laboratory tests did not reveal any membrane defects. The splits in the membrane were "V" shaped (in the split, the top plies were wider apart than the lower plies), showing that rotation was a featured element in the splitting. All of the components of the roofing system associated with the splits were wet.

The *designer* stated that the design was blameless, meeting all requirements of the local Building Code and good practice. The insulation–membrane combination met all the requirements for a Class A fire resistance.

The *roofing contractor* stated that he called the owner's attention to the insulation manufacturer objecting to the use of glass foam insulation on a steel deck, but he was told to proceed anyway, because they had to prepare the store for a major department store, a new tenant.

About two years later, we were again asked to investigate the roofing on the same New England department store. Our conflict-of-interest file warned of a previous assignment, but we accepted the new assignment when we realized that the new client was just a different department of our original insurance company client.

Our second field investigation was relatively short. Again we found splits in the membrane parallel to the bar joists. Roof cuts showed the new bottom-to-top construction was:

- the same steel deck,
- an asphalt-kraft paper vapor retarder hot mopped to the steel deck,
- a 76 mm (3 in.) thick layer of glass foam insulation,
- a 13 mm (½ in.) thick layer of gypsum board, and
- a gravel surfaced asphalt built-up roofing membrane.

The new splits took place where the butt joints of the gypsum board and the through joints of the glass foam insulation coincided.

QUESTIONS

1 What caused the original splits in the membrane?
2 Whose work was at fault (owner, design, workmanship, or other)?
3 What caused the second set of splits?
4 How can this kind of failure be avoided?

12 The case of the propriety products

United States government personnel consulted us about an unusual problem. Here is a summary of the facts as we know them through the benefit of hindsight.

A government-contracted architect prepared plans and specifications for re-roofing a government occupied building in the mid-west. The plans and specifications prepared required the contractor to:

- Remove the existing roofing system,
- Install a layer of expanded styrene (beadboard) insulation,
- Install a single ply self-adhering plastic film by Manufacturer A,
- Coat the film with a white coating by Manufacturer B, and
- The roofing to be installed by a contractor approved by the manufacturer of the roofing products (undefined further).

The contract documents were sent out for bid after approval by the local owner's representative. Almost as quickly as they went out, complaints by roofing contractors and local trade organizations echoed back. Maybe even some congressmen were involved. It seems only one minor contractor in the area was approved by the manufacturer, and the manufacturer refused to approve additional contractors without time-consuming training and credit checks. In addition, the products specified were propriety; they were called out by brand name without an "or equal" clause, contrary to the government's policy to avoid propriety products so as to foster competition.

The government quickly responded by calling back the bid documents and instructing the architect to modify the documents to comply with the government's anti-propriety policy, and to re-issue the revised plans and specifications at no additional cost to the government.

The architect took the simplest and lowest cost route. He added a few products from Sweet's Architectural File that sounded similar to the original specification, and an "or equal" clause.

After obtaining the contract, the lowest bidder roofing contractor attempted to purchase the roofing sheet originally specified from Manufacturer A. He obtained literature and a price list, but the Manufacturer again refused to sell him because they already had an approved contractor in the area, and the lowest bidder had not gone through the training the Manufacturer required. He then attempted to hire the approved contractor, but found that the approved contractor's price was too high; much higher than his low bid.

The lowest bidder then found a similar sheet product – although it had never been used on a roof – and submitted it for approval. At the same time, he submitted a coating product from a manufacturer that was willing to make a coating similar to, and lower in cost than, the Manufacturer's B coating originally specified. Manufacturer B advised the government that they would not sell him the coating specified. Both of these submittals were approved by the contracting officer without consulting the architect.

Leakage started shortly after the new roofing system was installed. The leakage area grew and intensified with each rain, and there was talk of abandoning the building.

Enter the roof consultant. When I visited the job I could view the logo of the sheet manufacturer through the almost transparent white top coating. There was no effective adhesion between the top coating and the plastic sheet. Indeed, some of the top coating had been displaced by the wind to litter the roofing surface.

The plastic sheet had shrunk to pull the roofing laps apart – enter the storm water.

Obviously the whole roof had to be replaced as soon as possible.

QUESTIONS

1   What actions or inactions contributed to the failure?
2   Which was the greatest contributor to the failure?
3   What should be done to avoid this kind of failure?

13 The case of the splitting membrane

We were called in to investigate the roofing system on a large warehouse. The building was steel framed. The roofing membrane was supported by a "Procure" gypsum deck supported by bar joists.

Poured gypsum decks have many advantages. Since gypsum gives off heat as it cures (an exothermic reaction), it can be applied in cold weather and can be roofed almost immediately. It qualifies as a fire resistant construction suitable for school, shopping malls, or other places of public assembly. Its sole disadvantage is that the resulting roofing surface is level (without slope), but even with this disadvantage, many very serviceable roofs have been constructed on gypsum decks.

Gypsum decks are installed by:

- Welding steel bulb-tees (similar to railroad rails in vertical section) to the top of the bar joists ~610 mm (~24 in.) apart, with welds on alternate sides of each bulb tee.
- Inserting fiberboard form boards so that they are supported by the flanges of the bulb tees.
- Draping square mesh wire reinforcement over the bulb tees (a special hexagonal mesh "chicken wire" with reinforcing strands going across the bulb tees is sometimes used).
- Pouring a ~50–100 mm (~2–4 in.) thick layer of wet gypsum, and screening the surface to level.

This construction is very forgiving. We had a case on one occasion where the reinforcing mesh in the gypsum was installed in the wrong direction. We load-tested the structure to full design load and found that all the deflection was in the bar joists; the gypsum deck did just fine.

Built-up roofing systems are typically mechanically fastened directly to the gypsum deck. An asphalt-coated base sheet is nailed to the deck with about 10 special fasteners per square metre (90–100 per roofing square) – older applications used cut flooring nails and metal caps. The balance of the roofing plies were then installed in a conventional manner.

In this particular case, long splits in the roof were found with their length parallel to the building's bar joists. Aside from the parallelism to the joists, the splits were at random, but we found they were always near the top of a joist.

Analysis of the splits showed the splits were "Λ" (an inverted vee – wider in the bottom ply than the top ply), showing the splitting force came from below.

Field investigation of the gypsum deck under the splits showed there were long linear cracks in the gypsum matching the splits in the membrane. The bulb tees butt joints fell on one bar joist, and there was an open sidelap in the reinforcing mesh sheets wherever the gypsum was cracked forming a vertical plane of weakness and stress concentration.

The deck installer stated that he always installed the gypsum the way he installed it on this job and never had any problems in his 20 years of experience.

Staggering the bulb tee butt joints on alternate or different bar joists was not required by the contract documents or the gypsum supplier. Forming side laps with the reinforcing mesh or tying the mesh sheets together periodically was not required by anyone.

QUESTIONS

1   Was the design work at fault?
2   Was the gypsum deck applicator at fault?
3   What responsibility did the gypsum supplier have?
4   What should be done to avoid similar problems?

14 The case of the unacceptable design advice

Several years ago we were employed by a local designer to help him with an assignment to develop a roofing design guide for facilities throughout the world. The assignment was from a department of the United States Government.

We were obviously elated to be chosen to do this work, and looked forward to be a positive influence in the production of reliable roofing systems of good quality, and to use the opportunity to educate a large number of designers on the peculiarities of roofing design.

We set to work enthusiastically to gather all the relevant information that was available about world-wide climates, peculiar local weather conditions and other economic and practical considerations that might be important to roofing design.

We drafted a broad but detailed set of recommendations, including a list of the roofing systems that were most likely to be serviceable in many of the climates, warnings about special conditions that might be involved due to occupancy of the facilities.

Our draft was submitted through our client to the government, where it languished for several months. We were then advised that: "Our draft had no merit and was completely unacceptable."

Obviously, this was quite a shock, and we attended a meeting with the government representatives, an architect from a central European state newly employed by our government, and representatives of our client.

The government's first objection was that we required the designer to go to the site location, to study the local conditions, materials available with the history of the materials and systems local performance, manpower available, typical weather limitations, and the facilities special use requirements. The government stated that they were not going to pay for the designer to "wander all over." They also suggested that only materials produced in the United States should be considered, and that installation could be done by roofers imported for the project – therefore no special local knowledge was needed.

The government's second objection was that we required the original designer to send his plans and specifications out for peer review by personnel skilled in roofing technology. The government's stated position was that they had contracted with the designer, and were not going to pay extra for his quality control.

The government's third objection was that we wanted the roofing work monitored by an independent party. The government stated that the contractor must be responsible for his work, and that monitoring didn't change the warranty. We agreed that monitoring did not change the contractor's duty to comply with the contract requirements, nor did it influence the warranty. These government representatives stated: "The warranty is the important thing!"

The newly hired architect from central Europe stated that it was apparent that we had little knowledge of weather, citing as an example, the potential problems with "salt fog" we mentioned in our draft. The architect said that it demonstrated a lack of knowledge of materials, because the salt is distilled out when moisture arises from the sea. He ignored the sections of US Army Corps of Engineers' documents discussing salt fog that we had quoted.

This situation demonstrates again that good advice is often ignored by the ignorant, increasing the probability of roofing failures, and the wastage of the tax payers' money.

QUESTIONS

1   How could this situation be handled? Seasons for hunting officious officials would probably be frowned upon.

> *Owner's competence*
> *Cannot be specified like*
> *Roofing materials.*

15 The case of the skilled maintenance man

Some of the "powers that be" in a very small New England town decided it was time to replace the existing roof on one of the minor buildings owned by the town. They consulted with several roofing contractors located in the neighboring communities (their town was too small to have a contractor) and were appalled by the prices quoted.

In their search for solutions to their problem, they discovered that the local hardware store also carried some roofing products. About a year and a half ago, the hardware store owner had started to handle a roofing sheet called: "Super Active Dreadnought Roofing;" he had not sold any, but was willing to give the town a special deal based on his civic duty (no one mentioned his penurious Yankee heritage). The literature said that the product was simple to install; it only required a large propane torch to melt the sheet in place, and offered on the roof training to anyone interested.

The Town searching committee read the product literature. They also found out that a local maintenance man had previously worked for a short time on a roofing crew (he was casual labor). In the committee's "wisdom" they decided that the maintenance man (who was also a volunteer fireman) and the Town Manager (who maintained his job by being a yes man) could easily install the new Super Active Dreadnought roofing – the literature said it required almost no effort.

The maintenance man and town manager borrowed a propane torch from the local plumber (who was also the building inspector) and forgave enough of the hardware stores taxes to pay for the roofing material.

The impromptu roofing crew decided to do the roofing work over a memorable Labor Day weekend. To their astonishment, they found that no one from the company that made the roofing sheet could be made available until at least two months in the future. They even had no one available on a national holiday weekend!

The crew decided to ignore the manufacturer's advice about cleaning off the existing roof (they had not borrowed a broom, shovel, or any other suitable equipment). They did manage to get the propane torch lit (one of the crew still smoked) and set about torching the polymer modified sheet roofing in place.

They had great difficulty melting the asphalt on the sheet, and the gravel from the old roof kept getting into their work area, but they persisted and finally covered about half the roof before they decided to quit for the day and have a few beers.

The maintenance man was roused from his evening meal by the town's fire alarm. The Town building he had been working on was burning. Later, it developed that the old punky fiberboard roof insulation ignited (from some mysterious source), smoldered, and the heat built up until a full blown conflagration consumed the building.

An experienced state fire marshal investigated the residue, and found that the fire originated on the roof in the area where the roofing crew was working. He opined that the fire had been caused by negligent roofing work by untrained personnel. He so testified in the court.

Some additional testimony revealed there had been some electrical problems such as electrical outlets that didn't work and fuses that blew every so often.

The jury heard that fires resulting from this type of roofing application were unfortunately quite common, and special fire protection methodologies (such as a fire watch, infra-red sensor scanning, and presence of fire extinguishers) were required, but not used in this case.

QUESTIONS

1   Whom did the jury find at fault?
2   What caused the fire?
3   What actions might have prevented the conflagration?

16 The case of the roof over the rare paper storage

A steel-framed building, with cement block walls, and a poured-in-place concrete roof deck was constructed to house valuable papers. The building was fire resistive throughout, included a sprinkler system, and a system to maintain the relative humidity at 50 percent to preserve the documents.

Shortly after the fall of the year, when the building was occupied, the inside face of the exterior walls began to darken near the ceiling as if they were moist. The general contractor called the roofing contractor because it was obvious that water was getting through the roofing system, traveling over the concrete deck, and coming out the deck-wall joint.

The roofing contractor investigated the condition of the roof, had his crew patch a number of suspicious locations, and reported that he had located the difficulty. It later was revealed that the principal of the roofing contractor company never went up on the roof; he was relying on reports from his foreman.

The dark stains became more pronounced and spread ever lower on the walls. Now, they were low enough so the wall could be touched without climbing a ladder. They were indeed wet to the touch. The technical representative of the roofing materials company that had furnished the bond or warrantee was called. He investigated the roof and reported that the roofing system installed met all of the manufacturer's recommendations and job specifications.

We were called in to investigate the causes of the leakage because many participants in the job feared a damaging law suit.

We started our investigation inside the building. The interior of the exterior walls were indeed wet despite a lack of local rainfall for over a week. They were wetter near the ceiling and the water was clean water, rather than the dirty water that evidences long travel or prolonged storage inside the roofing system.

We found that the roof was surrounded by parapets capped by precast concrete slabs. We made our first roof test cut near the northern parapet – where the water intrusion was the most severe. All components of the roofing system felt dry. The concrete deck under the roofing was dry. There was even undisturbed construction dust on the deck between the asphalt strip moppings that adhered the bottom layer of insulation to the deck.

A second test cut near another perimeter of the building yielded the same result – everything felt dry. We even found a piece of kraft wrapping paper from a roll of felt. The paper was dry and unstained by any previous water exposure.

We then examined the parapets. There was a metal through the wall flashing directly under the precast concrete caps. We could feel air almost whistling out from underneath the flashing. The air was moving fast enough to blow out lit matches held near the opening.

QUESTIONS

1   What caused the water intrusion?
2   Whose work was at fault?
3   How can this problem be prevented?

17 The case of the wide spread flame

Very early on a blustery morning a fire broke through the roof of a manu-facturing plant in the Midwestern United States. The entire building was involved, and eventually burnt to the ground.

The factory was metal-framed and sided. The roofing system consisted, bottom-to-top of a steel deck, a minimal quantity of fiberboard insulation, and an asphalt built-up roof. A few years before the fire, a new roofing system, consisting of a polyisocyanurate foam recover board and a fully adhered EPDM membrane were added. The building did not have a sprinkler system, but was fully alarmed.

The building was occupied by a company that bent metal into various shapes required by other industries. Typical operations included massive metal shears, hydraulic bending and stamping machines, oil treatment baths, and other similar equipment for forming metal. The building was heated by a gas fired hot air heater fastened to the underside of the steel deck that was exposed to the interior of the building.

The owner's insurance company paid off after the fire, and attempted to recover from other parties. Sued were the manufacturer of the heating equipment, the fire alarm company, the roofer that installed the remedial insulation and roofing, and the manufacturer of the remedial roofing insulation and roofing.

Here are some of the various theory of liability used by the insurance company, and the responses by the defendants: Based on the fact that the fire started near the space heater, the space heater manufacturer was blamed for defective equipment. His response was that the space heater was designed to fail safe – shut down in any emergency, including power failure. Also, there was never any evidence presented that the space heater was defective in any way.

Delay in reporting the fire permitted it to get out of control, is the basis for the case against the alarm company that installed and monitored the fire alarms. It was argued that the majority of the building could have been saved, had the alarm been sounded sooner. Evidence showed that the fire started near the space heater on the western side of the building and spread to the eastern side before the fire alarm was sounded. Firemen responded to the alarm and were at the building within five minutes of the alarm, but the fire had already involved the whole building and burned through the roof. The alarm company's defense was that the alarm system was correctly installed and passed an inspection just a few months before the fire.

The insurance company relied on two events for their charges against the roofing contractor and the remedial roofing materials manufacturer. These were:

- A well-known book on roofing design stated, in part, that installing a new roofing system over an existing roof can increase the building's fire hazard, and
- Many years before, the Livonia, Michigan fire in the General Motors Plant demonstrated that a fire can be quickly spread by burning asphalt.

The insurance company's charge was that since the additional roofing system increased the fire hazard for the building, and the fire was quickly spread by the Livonia Effect, and both the roofing contractor and the roofing materials manufacturer failed to warn the owner of the increased hazard, they were responsible for the total destruction of the building. "Failure to warn" was the principal charge.

The roofing materials manufacturer responded that:

- The building prior to re-roofing was a "Class 2" construction (without a sprinkler system). While it is true that a Class 1 building may be classified as Class 2 due to the addition of an unapproved roofing system, a Class 2 building remains a Class 2 building regardless of additions. There is no Class 3 of fire resistance. Therefore the fire hazard classification was unchanged by the addition of the second roofing system. Since there was no change in the fire hazard classification, there was no duty to warn by anyone.
- Thermal calculations show that the heat of the fire under a steel deck is determined by the fuel feeding the fire, not by the quantity of roof insulation above the steel deck.
- Testimony was offered to suggest another conflagration path. In most manufacturing facilities it is quite normal for an oily condensate to be deposited on the underside of the steel deck. This becomes contaminated with dust to form oily lint that can easily flash spread a fire. The existence of such an oily film is supported by testimony about large open oil baths on the floor of the facility. A flash fire can quickly spread a fire, too quickly to be detected by a fire alarm, until a larger, hotter conflagration develops.
- No evidence was offered to explain the source of ignition.

QUESTIONS

1   What ignition source started the fire?
2   How was the fire spread?
3   What responsibility did the roofing contractor have for the loss of the building?
4   What responsibility did the roofing materials manufacturer have for the fire loss of the building?
5   What could have been done to prevent this conflagration?
6   Assume the case was settled. Who paid the largest share?

18 The case of the dissolving shakes

The Arbuckles, both senior citizens, live in the forested area of western Oregon. About four years ago, they decided to have the shingles on their twelve year old home replaced. They considered the following options:

- asphalt shingles (which they were told would last about fifteen years),
- wood shakes (that might last about ten years and were not fire-resistant), and
- artificial shakes (fire resistant and guaranteed for 50 years).

The artificial shakes were made from Portland cement and reinforced with wood fiber. They were said to be rot resistant, and not harmed by freeze–thaw cycles. The Arbuckles had their roof replaced with the artificial shakes based on these promises. While these shakes cost more than the other alternatives, Mr and Mrs Arbuckle had just come into a bit of money, and they decided to spend the money so that they did not have to replace a roof again in their lifetime.

About two years after they had the new shakes installed, Mrs Arbuckle noticed that some of the coating was coming off the shakes. Mr Arbuckle investigated and found that the coating on the shakes was starting to peel off on the south side of the house. On the north side, he found that moss and fungus were growing out of the bottom edges of the shakes. Mr Arbuckle then called the installer, who referred him to the manufacturer of the shakes. It took several telephone calls to get through to the Customer Service Department who finally agreed to have the local representative look at his roof.

The manufacturer's representative viewed the roof from the ground, and reported that the only problem he saw was that the installer failed to use the starter shingles along the eaves of the house. The flaking coating was normal and could be easily taken care of with another coat of paint, supplied free by the manufacturer (the Arbuckles would have to pay for the application labor), and the moss and fungus growing out of the shingles on the north face was due to the Arbuckles' poor maintenance. They should have kept the roof clean.

Obviously the Arbuckles were incensed, and felt that they had been victimized. They contacted the Western Roofing Contractors Association and found out that cement artificial slates and shakes, reinforced with cellulose (wood) or other water absorptive materials, had been the source of many complaints. They also found out that the fake shakes were no longer in production.

The manufacturer stated that there wasn't a problem since no leaks into the house had yet been reported. They thought it was reasonable to maintain the roof by replacing the paint coating every few years, and to keep the roof clean of anything that might land on it. Their expert opined that there was nothing wrong with the roof, except the missing starter shakes, but would not say that the roof would last 50 years. There was no connection established between the missing starter shakes and the flaking paint or the moss growth.

QUESTIONS

1 Has failure occurred?
2 If failure has occurred, who is responsible?
3 How could this situation be avoided?

19 The case of the leaking gymnasium

Gymnasiums seem to draw more than their share of roofing leaks. Some say it is because the expensive wood floors attract the water – I doubt it, but probably the relatively long span required to roof several basketball courts has some influence on leakage. In this case the roof of the gymnasium was supported by long precast concrete double tees. There were stair wells jutting off two apposing corners of the gym; the stair wells are capped with cast-in-place concrete decks.

Shortly after the first winter, leakage was reported along the masonry walls supporting the precast tees, and on the walls between the gym and the stair wells (ice seldom leaks into buildings). The designer asked us to investigate the leakage.

We confirmed the locations of the leakage into the interior together with the designer, and went up on the roof with the roofing contractor. Getting up on the roof was quite a problem because the school did not have ladders long enough to reach the top of the parapet. We finally managed to get on the roof with a special extension ladder.

The precast double tees were properly cambered (arched slightly upward in the center of the span to allow for creep and deflection); the designer used the camber as a slope to direct water toward the roofs over the stairwells that contained the roof drains. Both drain strainers were partly clogged with leaves and other debris. Water ponded near the drains ran freely down the leaders as soon as we lifted the strainers.

We saw splits in the roofing directly over the intersections between the cast-in-place concrete over the stair wells and the precast tees over the gym. The parapet flashing was torn over the ends of the precast tees, and along the crest of the cambered tees.

Storm water, in order to leave the roof, was directed toward the ends of the tees by the camber. It then had to cross over the joint between the tees and the cast-in-place deck where the drains were located. Any differential movement between the structural systems would and did rupture the roofing.

Differential movement was also shown by the tears and displaced parapet flashing caused by the downward deflection of the double tees, and by the planks end rotation. Most of the storm water drained off the cambered tees to the low ends, entered the torn flashing to leak into the building.

Some of the storm water that moved toward the drain would pour into the splits in the roofing. This leakage was made more serious by the clogged drain strainers that prevented the free flow of water down the drains.

QUESTIONS

1 What work (design, workmanship, materials) contributed to the problem?
2 What features of the design might have avoided the problem?
3 Did the roofing contractor have any responsibility for the leakage?
4 What should be done to avoid this problem?
5 What was the designer's responsibility in regard to providing easy access to the roof for maintenance?

20 The case of the flapping glass fiber felts

We receive many telephone calls from people with problems. Fortunately, we can often address the caller's concerns using our long experience, but occasionally we hear about a new strange problem.

Such was the case one summer day, when the representative of one of our former clients called to report a very unusual incident. He was monitoring re-roofing work at the Midwestern factory where he was the facilities manager. On the day before his call, a private roofing contractor had removed a large segment of the existing roof and insulation, and installed two layers of new insulation and four glass fiber-felt plies in hot coal-tar pitch.

The caller had been on the job all day, and everything went very well despite the unseasonably warm weather. They elected not to finish the work on this roof segment with the specified coal-tar pitch flood coating and gravel surfacing because night was falling as they finished the four ply – they had attempted to remove and replace too large an area for the work crew at hand.

When the caller went back to the roof the following morning, he found loose glass fiber felts flapping about in the wind all over the roof. That's when he decided to give us a telephone call. We agreed that this was a new event in our experience, and we asked him to send us samples of the felts by overnight mail – (we had some of the coal-tar pitch in hand).

The felt samples were very "open." If you laid a felt ply on a newspaper, you could read the paper through the felt. We measured the absolute density of the felt and the coal-tar pitch using isopropyl alcohol – (the alcohol's low viscosity permits a quick and accurate determination). We found that the asphalt-coated glass fiber felt had a much lower density than the pitch – hence the felts floated on the pitch. The movement of the felt plies from the center of the membrane to the top was aided by the hot weather, the high porosity of the felts, the low viscosity of the coal-tar pitch, and the stiffness of the glass fiber felts that "remembered" their rolled state prior to application.

Since this incident, and others like it, work is ongoing in ASTM International's Committee D08 on Roofing and Waterproofing to develop a test method to measure the porosity of roofing felts, and criteria to limit felts' porosity. Progress has been slow because it is difficult to gain consensus on the many test variables, because the problem currently seldom occurs, and because some of the representatives just don't want a standard.

QUESTIONS

1 Where does the fault lie in this instance?
2 Should the roofing materials manufacturer have foreseen this eventuality and warned about leaving membranes out without a flood coating and gravel?
3 Considering the long history of asphalt and pitch incompatibility, why were asphalt and coal tar products specified in contact with each other?
4 What was the obligation of the roofing contractor?
5 Given the specialized knowledge of the materials manufacturer, what was his responsibility?
6 How should the roof be fixed?
7 Who should pay for the remedial work?

21 The case of the leaking condominiums

Condominiums present a host of problems – some engineers will not even take them on as a client because of sad prior experiences. Condominium associations have their own problems with rotating members on the board of owners; many of these owners have little or no experience in construction or property management. They tend to rely on the property manager they hire, who sometimes knows less than the owners.

Some of the problems are due to the developers, who are trying to use the economies of scale to produce relatively economic housing – they, like building materials manufacturers, tend to continue to improve their product until it does not work – but, in their defense, if there is a fundamental design error, it is multiplied by the number of units involved. A lawsuit is predictable just as soon as the ownership is transferred from the developer to the Condominium Association. This is aided by some lawyers who make condominium disputes their specialty.

Such is the case in point. A lawyer had obtained a list of the applications of a roofing system by a building materials manufacturer through court orders in a previous case. He used an engineering firm (who always seemed to be his alleged expert) to explain to the Condominium Association that the roofs were defective, the building materials manufacturer had deep pockets, and they should protect themselves by hiring this well-known attorney. Guess who?

Defending a case like this that involved over 300 roofs is not simple. How do you prove that none of the 300 roofs leak? Just saying they don't leak doesn't cut it. Maybe some of them leak, but the leakage may not be due to the roofing system employed.

In this case we used a statistical fact. If you have a 50–50 situation; heads or tails; male or female; on or off; or leakage through the roof or none; if you select and examine 23 random samples and find all of them to have experienced leaks through the roof, or all of them not to have experienced leaks, you can be more than 90 percent confident that you have determined the truth about the entire population under study. This is true of every population size, within reasonable limits. Ninety percent confidence is more than is required by any professional testifying.

We did such a study by examining and testing samples from 23 roofs chosen at random. In addition we examined three roofs that were supposed to demonstrate catastrophic leakage, and didn't find any leakage through the roofs. We did find some leakage through uncapped parapets, open television antenna and air conditioner penetrations that were not related to the roofer or building materials manufacturer. These data were used to suggest that the building materials manufacturer had no duty to make a contribution for the leakages alleged.

QUESTIONS

1 What do you think about the actions of the engineer friend of the lawyer?
2 What do you think about the actions of the lawyer?
3 How can you avoid this kind of difficulty?
4 How was the case settled?

22 The case of the corroding foam

A new office park was constructed on the outskirts of a major city on the Eastern coast of the United States. The buildings were metal framed, and the roofs were supported by steel decks. The roofing system consisted of a phenolic foam insulation and a multiply asphalt-glass fiber felt built-up roofing membrane.

Shortly after the built-up roofing was installed, iron workers installed a metal frame and a sloped plywood deck about the perimeter of the roof to form the look of a mansard roof. The plywood was roofed with asphalt shingles.

The interior was built out for the tenants as they signed leases. Among other things, sanitary vent penetrations had to be cut in the new roofing for the new bathrooms, and the original roofer was assigned the task of cutting and flashing these penetrations.

The roofing mechanic who made the roof cuts was shocked to see that the steel deck was severely corroded. He reported the condition to his boss, who informed the General Contractor in turn. The owner brought in the roofing consultant, and his investigation showed that the rusting of the deck varied between severe to catastrophic (where the deck had been rusted through to the interior). Small deck areas near the drains were free of rust where tapered perlite insulation had been used to form the drain sumps. The built-up roofing had holes in many places near the fake mansards that was consistent with damage during construction.

Eventually, the whole low-sloped roofing system (both the membrane and insulation) was removed; the deck was remediated by wire brushing and paint, or by replacement, and a new roofing system was installed.

The *roofing contractor* privately blamed the general contractor for not protecting his work when the mansard system was installed by another sub-contractor. He was very quiet about his thoughts because he depended on the general contractor for a lot of contracts.

The *insulation supplier* blamed the general contractor for failing to protect the built-up roof. The deck would not have corroded if the built-up roof had been watertight. In addition, he blamed the designer because he did not specify a vapor retarder that was required by the Building Code.

The supplier reasoned that a good vapor retarder would have kept the acidic water away from the roof deck.

The *mansard contractor* blamed the design of the low-sloped roofing, saying it was too soft to survive as a working platform, which the design required. He also suggested that the general contractor pushed the work on the job too much – such that a good workmanlike job was impossible.

The *general contractor* blamed the corrosion on the selection of phenolic foam insulation, which was inherently unsuitable for use, because any roof insulation can expect to get wet sometime in its life, and must survive that exposure. In addition, the mansard contractor was at fault because he did not perform his work when it was planned, before the built-up roof was installed.

The *designer* blamed the insulation manufacturer for not warning of the severe consequences of getting the phenolic foam wet. Some reports stated that formic acid was the active material exuding from the wet foam. Formic acid is the "poison" in a bee and other insect stings. This raised the question about consequential damages to individuals with allergies to insect bites. In addition, the designer blamed the general contractor for not controlling the proper sequence of the work.

QUESTIONS

1   Considering the work of the designer, insulation supplier, general contractor, roofing contractor, and mansard contractor – how much fault belongs to each? Why?
2   Was the "vapor retarder" argument offered valid?
3   What was the real world outcome? Who paid for what?
4   How could this problem have been avoided?

23 The case of the heavy glass fiber shingles

A designer of condominiums specified a 265 lb (120 kg) per sales square three-tab strip shingles for a large number of condominium units located in the north-central United States. After-the-fact hints indicate that the designer wanted asphalt shingles based on organic felt, because of their history of satisfactory service.

One potential supplier manufactured asphalt-glass fiber felt-based shingles, felt he could meet the weight by increasing the quantity of filled asphalt coating in his standard three-tab asphalt shingle formulation. Both the supplier and applicator were creative in their bidding, and were awarded the contract.

The shingles began to split within a few years of installation, and many units reported leakage. Investigation showed that the shingles did not meet the minimum requirements of ASTM International's Standard D3462 on asphalt-glass fiber shingles in tear and pull-through strength. Shingles from the roof have average Elmendorf tear strengths of 900–1200 gm (the Standard requires a minimum average of 1700 gm). The Condominium Association elected to sue the shingle supplier for providing a product unsuitable for use. They wanted all the roofs to be replaced at no cost to the owners.

The *designer* agreed that the shingles supplied met with the written specifications, but relied on the special expertise of the manufacturer to provide shingles that were suitable for use. From his vantage point, the shingles failed to meet the minimum requirements of the ASTM International Specification, and split in service.

The *shingle supplier* responded that he had no duty to meet ASTM specifications since compliance was voluntary. In addition, they argued:

- The Standard only applies to newly manufactured product, not to product that had been exposed to the weather. Elmendorf values were known to decline with both age and exposure. The lower values measured were on shingles that had been exposed to the elements, and this did not prove that the shingles did not comply when they were first manufactured.

- The Elmendorf test is inappropriate because it provides a twist or out-of-plane force that does not exist in applied shingles; only a tensile test is valid. In any event, the Elmendorf test is quite inaccurate; their shingles would pass if the normal test variation is applied.
- The pull-through test is not appropriate in this case because splitting, not fastener pull through is the failure mode.
- All they have to do is repair a few torn shingles.

The *Condominium Association* representative responded that compliance with the ASTM Standard D3462 was a requirement of the Building Code. Standard D3462 does apply because there is no standard that applies to weathered shingles, and this consensus standard, which the shingle supplier helped develop, is all that is available.

Tests by others show that the Elmendorf tear strength has a much better correlation to the shingle toughness and resistance to crack propagation than any other property, including tensile strength. The minimum value cited is for the average of ten determinations. Compliance requires the average to meet or exceed the 1700 gm value, without applying any tolerances. The effectiveness of the Elmendorf test can also be demonstrated by the fact that products made with organic felts that have high tear strengths never have been known to split, while products made on glass fiber felts with low tear strengths frequently split.

Replacing only the torn shingles is impractical because all the shingles are firmly adhered to each other.

QUESTIONS

1 Which party is at fault?
2 What remedial action is appropriate?
3 How can this problem be avoided?

24 The case of the moving insulation

Quite a few years ago, a scourge of roofing failures descended on a Canadian province. Almost every new roof on a school was affected. The symptoms were widely dispersed splits in the roofing membrane, and severe, wide spread, leakage.

Most of the schools had a poured-in-place steel reinforced concrete deck, one or two layers of extruded polystyrene insulation, an asphalt-coated organic felt base ply, three asphalt-saturated organic felt plies, and an asphalt flood coating with a gravel surfacing.

The supplier of the extruded foam lauded the water resistance of the foam, pointing out that it was used for flotation in many applications, and stated that a vapor barrier or vapor retarder was unnecessary.

Careful installation is required because of the low deformation temperature of the foam – 60 °C (140 °F). For example: after mopping hot asphalt on the deck, an installer would drag a corner of the insulation board in the asphalt to be sure it was not too hot. The board was quickly installed if the corner did not melt, before the asphalt became too cool to obtain adhesion. The base felt would usually be installed by the "mop-and-flop" method, where the underside of a short run of basesheet would be mopped with hot asphalt, and then flipped into place on the insulation – again, before the asphalt cooled too much.

At one point in time, a supplier provided a "self-adhering" base sheet. The base sheet was applied without interply asphalt; it used the heat from succeeding moppings to soften the asphalt on the back side and to adhere the base sheet to the insulation.

Field investigations of these split roofs revealed:

- The roofing membranes were very wet; 10–20 mass percent, based on the dry membrane weight, were not unusual determinations.
- The extruded foam insulation was very wet; some measurements showed that the water present weighed three or four times the dry insulation mass.
- Splits in the membrane were invariably over joints in the insulation boards.

- Most splits were parallel to the length of the felts in the membrane, but a few turned corners to match the insulation joints below.
- The membrane splits were Λ shaped; wider at the bottom of the membrane than at the top; illustrating that the splitting force came from below the membrane, and that the membrane split over a period of time rather than a sudden catastrophic fracture.
- There were no cracks in the structural decks in the vicinity of the membrane splits.
- We often found insulation joints gaping as much as 13 mm (½ in.).
- The attachment between the membrane and insulation varied. It was excellent near the splits and almost non-existent in other areas. The insulation to deck attachment varied considerably.

The *designers* were universally blamed for improper specifications. The problem was so severe that designers throughout the province were threatened with loss of their professional liability insurance.

The *insulation suppliers* blamed the roofers. Curiously, when the problem became evident in the United States, the supplier blamed poor application, and increased his marketing efforts in Canada, where he believed the workmanship was superior. Now that the problem repeated in Canada, the supplier increased his efforts in Europe, where he believed the workmanship was better. Eventually we investigated similar split roofs in France and Germany.

Studies showed that while the insulation might not need a vapor retarder (a questionable assertion), the roofing membrane and system surely needed that protection. Moisture driven from the warm concrete, through the joints in the insulation, to the cool membrane, condensed to wet the system and weaken the membranes' felts. Insulation mobility and shrinkage provided the splitting force. This is confirmed by the evidence showing the force for the splits came from under the membrane and above the deck. Extruded polystyrene was the only material in that location.

QUESTIONS

1   What was the splitting mechanism?
2   How could this splitting problem be avoided?

25 The case of the phantom deck movement

A Superintendent of Schools in a New England community called because a historic building at one of his schools that had been recently reroofed, was leaking. He had first called his roofer, who advised that the shingles were splitting, letting in the storm water.

The roofer contacted the shingle supplier, and a local salesman viewed the roofs involved. After some short investigation, the local salesman announced that the splitting of shingles was due to movement of the structural deck, and failure to properly vent the attic space directly under the shingles. These conditions were specifically excluded in the supplier's warrantee.

We responded to the Superintendent's call and toured the building with him. The building involved is a relatively long narrow building, with its major axis east–west. As with many older buildings, we found that this old charmer is actually three buildings joined together. They all have steeply sloping decks, and each has a separate and different structural system.

The first deck is composed of double tongue and groove lumber 64×140 mm (nominal 3×5 in.) planking installed vertically up the roof. Drying shrinkage had opened some of the joints between the aged planking. The second deck is sheathed with plywood and the attic area below is not vented. The third deck is plywood over an area used as a garage; it is not heated or insulated.

We estimated the proportion of shingle splits in each roof area; there were fewer splits on the roofs facing north than the roofs facing south. We could see no significant difference in the splitting intensity in the shingles over the unheated garage, the heated and unvented area and the heated plank deck area.

We collected samples from each of the roof deck areas. In every case, we found the #15 shingle underlayment unbroken. Below the felt, there was no connection between any feature of the deck and the splits in the shingles.

We analyzed the shingles recovered during our field investigation, and measured the tear resistance. The weight of glass fiber felt was close to the minimum specified in ASTM International D3462. The tear strength averages 1200 gm, which is consistent with the manufacturer's specifications, but below the 1700 gm minimum in the ASTM standard. The quantities of asphalt, granules, back surfacing, and filler are consistent with the supplier's specifications.

The *supplier* stated that they designed the shingle to be competitive, and that they could not afford to make this shingle comply with ASTM Standard D3462 because they would lose market position.

QUESTIONS

1   What do you think of the idea that a supplier must make a product of less than minimum quality to maintain market position?
2   How can this kind of situation be avoided?
3   How was this problem settled?

26 The case of the lightweight insulating concrete

There are several kinds of lightweight concrete, including structural lightweight (using a strong lightweight aggregate), cellular lightweight (concrete foamed with gas), and vermiculite lightweight concrete (using vermiculite aggregate). This case is about vermiculite concrete.

Vermiculite is a mineral based on mica that greatly expands when heated. It is used in agriculture to loosen soils rich in clay and to retard drainage by holding water by adsorption on its broad surface area. In concrete, vermiculite aggregate reduces the density of the concrete, increases the thermal resistance, and tends to hold water for prolonged intervals; it retards drying.

An owner asked us to investigate the roofing system on a corporate headquarters in the New England states. The specifications for this new building required a poured-in-place concrete deck with a vermiculite lightweight concrete and expanded polystyrene fill to provide insulation and to slope the roofing surface to interior drains. A built-up roofing membrane was to be applied by nailing an asphalt-coated base sheet to the deck with special fasteners, hot mopping three asphalt-organic felt plies, and flood coating the surface with hot asphalt to adhere gravel surfacing.

Our investigations revealed that water was entering the building around storm drains and other deck penetrations. The roof was marked by patch repairs of splits in the roofing membrane parallel to the felt direction, and many, widely dispersed, blisters. There were many blisters in the built-up wall flashing, and many of the roofing membrane and flashing blisters were broken.

Our field sample cuts showed that everything was wet. The vermiculite concrete contained more water than it contained when it was poured. Hand pressure squeezed water from the expanded polystyrene insulation. The blisters were between the base and top ply sheets. Water almost always seemed to be the dominant constituent, no matter where we cut the membrane.

Laboratory analyses confirmed that all components were very wet. Solvent extraction of the roofing membrane to remove the asphalt present showed the felt plies were deteriorated; they were like soft punk. The quantities of asphalt in the membrane were appropriate, as were the quantities of adhered and loose gravel surfacing. The number of fasteners ranged from 8 to 9 fasteners per square metre (90–100 fasteners per square).

Eventually, all parties agreed that the fundamental problem was water built into the roofing system at the time of construction. Since the volume between the concrete deck and the membrane was sealed, the water in the system could not escape; it rather attacked the roofing membrane, weakening the felts, and inflating blisters wherever a void was present. The weakened membrane and blisters split, and storm water added to the water in the system. The water the system could not hold leaked into the building through every unsealed deck penetration.

The *designer* stated that he had followed the advice of the insulation supplier and the roofing materials supplier. The expanded polystyrene insulation layer below and integral with the vermiculite concrete was supposed to drain and vent the concrete. He examined the concrete before the roofing work was started and reported that the concrete was slightly damp and fully cured.

The *insulation materials supplier* agreed that the polystyrene would act as a venting layer, but only if the perimeter and penetration flashing was vented and permitted the system to dry out.

The *roofing materials supplier* showed that their literature warned against and excluded from warranty any application over vermiculite lightweight concrete. They also opined that without a positive flow of air through the system, it would be many years before the system would dry out.

QUESTIONS

1 Who was at fault?
2 What remedial action was appropriate?
3 How could this disaster have been avoided?

27 The case of the shrinking insulation

We were asked to investigate the roof on a "cooler" building (where the ambient interior temperature is kept at or below 4 °C [40 °F]) that contained food for distribution to local super markets. During the winter months, the tenants complained of almost constant leakage through the roof and sporadic leakage (during or shortly after rainstorms) the rest of the year. Leakage is a very serious problem for this food distributor, since the Board of Health could force them to cease operations until the leakage was halted.

The roofing system was supported by steel bar joists and a steel deck. The specifications required multiple layers of "iso" (polyisocyanurate foam) insulation and a fully adhered EPDM roofing membrane.

Not much was revealed by our inspection of the interior of the building. Dark spots on the concrete floor were widely spread, and the pallet racks and stored products masked any signs of active leakage. We did not see any water dripping off the bar joists, but stains, consistent with water leakage stains, were evident in several areas.

Up on the roof, we saw a smooth black plain that is typical of a fully adhered EPDM installation, except for shallow depressions like gutters scattered randomly over the roof. At some of these "gutters" we found open seams in the EPDM, with air from inside the building blowing out through the open seams. The underside of the EPDM was wet over the gaping insulation joint at every location checked.

At test cuts, we found a gap of up to 100 mm (4 in.) between adjoining insulation boards in the top layer, and somewhat smaller gaps between boards in the lower layer. We cut strips of EPDM out of the membrane to measure the size of the insulation panels. The nominal 1.22×2.44 m (4×8 ft) panels actually measured 1.19×2.40 m (47×94⅝ in.); an apparent shrinkage of 25×37 mm (1×1⅜ in.), or 2.1×1.43 percent.

Eventually all parties agreed that the "leakage" was in part condensation, caused by the moist interior air, channeled by the gaps in the insulation and the areas where the EPDM's seams were open, impacting the cool underside of the membrane during the winter months. The moisture condensed on the underside of the membrane and the condensate eventually dripped down into the interior of the building. The rest of the leakage was due to rainwater gathered by the "gutters" and channeled through the open seams into the building.

The *roofing contractor* claimed his work was blameless; pointing out that he just would not have left those gaps in the insulation. He opined that air from the pressurized building opened the membrane laps before they had time to cure. He complained about being pressured to install the roof during questionable weather. He suggested cutting out the "gutters," installing new strips of insulation to fill the gaps in the top layer, and installing EPDM strips to seal the openings.

The *insulation supplier* claimed that the dimensions of the insulation boards were within normal factory tolerances. He suggested that the open membrane laps were caused by the roofer attempting to seal the laps over the gaps that he left between the insulation panels. He also complained that the construction work was a fast track job, and his organization was unable to check the job the way they normally would with conventional construction times.

The *general contractor* agreed that the job was partly on a fast track due to the late shipment of steel for the structure, but stated that was done with everyone's knowledge. He asked the insulation supplier about reports he had of new insulation shrinkage caused by inadequate aging prior to trimming at the factory (he never received an answer).

QUESTIONS

1   Who's work was at fault?
2   What is your opinion about the fix suggested by the roofer?
3   What probably could have prevented this problem?

28 The case of the gooey felts

A New England school had a puzzling problem. Their new school leaked every spring. The leakage was widespread and seemed to moderate slightly during summer. The roofing contractor involved, spent many hours searching for holes in the membrane. In desperation, he removed the loose gravel surfacing and flood coated entire roof areas with hot coal-tar pitch. This usually stopped the leakage – until the next spring.

As specified, the bottom-to-top roofing system was a dead level concrete deck, two layers of insulation, and a four ply asphalt-glass felt membrane in coal-tar pitch, and a coal-tar pitch flood coating protected by a white marble aggregate. A central area, over the library was to be sloped at 17 percent (2 in./ft); it was to have the same roof, but with steep asphalt as the bitumen.

We were called in by the roofing contractor, who could not find or stop the leakage. We had our first view of these roofs on a hot summer day, while the roofing contractor was flood coating another section of the roof in an effort to stop the leakage. The sloped asphalt roof was brilliant white – the other roof areas were black. We saw pockets of a greasy material approximately 1 m (3 ft) in diameter here and there over the surface of the pitch roof. The white gravel was severely stained, as if by fuel oil.

We sampled every roof area. The only membrane in good condition was the asphalt membrane that was over the only roof area that had never leaked. We were amazed to see how easily the gravel and flood coating was removed at the perimeter of the pitch membrane samples. (Pitch is normally very hard to remove off membrane – dry ice is sometimes required to remove the flood coating off a pitch membrane.) Water drained from the glass fiber felt membrane when we lifted the sample. The felts were easily separated from each other. All components felt wet.

Laboratory examination of the samples from our field investigation showed the asphalt membrane was dry and contained appropriate quantities of the specified material. The pitch membranes were all very wet. The average quantity of interply pitch was very low – it was lower than we know a contractor can apply in the field – and macroscopic examination showed that the interply pitch present was in bundles of strands similar to spaghetti.

The flood coating quantity was several times higher than we would expect to find on a coal-tar pitch roof.

The *designer* stated that he followed all the recommendations of the roofing materials provider.

The *roofing contractor* stated that the roofing was installed as required by the contract documents. Despite his best efforts, the weight of the gravel surfacing pressed the glass fiber felts down through the membrane, straining the coal-tar pitch through the felt plies, where the pitch picked up some of the asphalt from the asphalt-coated glass fiber felts, and an incompatible reaction resulted that destroyed the structure of the pitch so that it was no longer waterproof.

The *roofing materials supplier* said he had every assurance from the asphalt-glass fiber felt supplier (a different company) that many roofs of this type had been installed without any problems.

The *school board* demanded a completely new roof after the existing roof was removed.

QUESTIONS

1 Who was at fault? Why?
2 What is your opinion about the school board's request?
3 How could this problem have been avoided?

29 The case of the ice castles

Freezer buildings pose many problems – not the least of which is that a vapor barrier is required. A vapor retarder is seldom enough. Experience suggests that the best plan is to build a building within the building to house the freezer, but the importance of economics and the growing size of the freezers to serve the market make the best plan impractical. Modern prefabricated insulated panels make the design of stand-alone freezers attractive.

We were called to a freezer because roof leaks were creating icebergs, threatening to topple the walls. The roof deck consisted of insulated panels, as previously mentioned, with plywood faces and expanded polystyrene foam cores. The panels were covered with a relatively thin layer of glass fiber insulation and a smooth asphalt-asbestos felt membrane.

Inside the freezer, we saw quite large chunks of ice adhered to the walls at the wall–ceiling juncture. We saw no other evidence of leakage. The outside walls of the building were sheathed with corrugated asbestos-cement panels.

We carefully examined the surface of the roofing membrane and found no obvious dislocations, splits, or blisters. The roof sloped slightly to a perimeter gutter; it had no penetrations that we could see.

We cut several samples in the field (the central area) of the roof. All components were dry to the touch, and the plywood skin of the roof panels showed no evidence of being exposed to water at any time.

We cut some samples at the perimeter of the roof. The glass-fiber insulation was water logged; water drained from it as we lifted each sample. We found a relatively thick plastic sheet on the plywood under the roof insulation that terminated about 300 mm (1 ft) from the edge of the roof. We came to realize that the plastic sheet was the vapor barrier from the wall system that was placed over the plywood panel. The roof insulation and membrane were installed without making a sealed connection between the roofing membrane and the vapor barrier from the wall system. Moist air moved up the corrugations in the exterior panels by a chimney effect to be directed into the glass-fiber insulation by the perimeter flashing for the roof. A less than perfect joint between the wall and ceiling insulation panels permitted a cold spot to develop where the moisture in the air condensed to drip into and freeze in the interior of the freezer.

The *designer* blamed the roofing contractor for not making a seal between the roofing membrane and the wall's vapor barrier. He also blamed the general contractor for not joining the ceiling and wall panels completely (a rough joint was found covered with a wooden batten), and hiding his defective work.

The *general contractor* responded that his crews did the best they could joining the wall and ceiling panels. The dimensions of the work as designed required cutting the insulated panels; these cuts of the thick insulated panels were very difficult in the field.

The *roofing contractor* responded that he installed the roof precisely as it was designed, and asked the designer to show him where the contract documents required him to attach the roofing membrane to the wall's vapor barrier (an examination showed the perimeter flashing detail was drawn as part of an overall vertical section through the wall, and the designer's intent was unclear).

QUESTIONS

1   Who was at fault?
2   What actions might have prevented this problem?
3   What lessons can be learned from this failure?

30 The case of the tile

Ceramic tile roofs have been in use for centuries. The tiles used on many European castles during the Middle Ages and the Renaissance; these tiles were often overlapped with wooden elements. All that held them in place was often their own mass and little knobs on the rear of the tiles that were hooked over the wooden supports. Even earlier, Chinese pagoda roofs were massive assemblies of wood and tile. They relied on the heavy weight of the tile roof to stabilize the building during windstorms. Without secondary protection, I suspect the ancients were more tolerant of leakage than modern man, because I'm sure many of these roofs leaked, even with the very steep-slope upon which they were installed.

The modern tile roof part of the roofing industry has quite a bit of internal conflict between producers who strive to make a dense, low water absorption, strong, long lasting material that can survive freeze-thaw exposures, and producers that want to make lighter in weight, strong tiles, that probably will not do well in freeze-thaw, but are much more economical because of their lower mass and lesser structural support they need.

Most of the tiles are sold in warm climate areas, and unlike the ancient castle roof's 100 percent slope (12 in./ft), the modern roof may be lucky to have a 17 percent (2 in./ft) slope. A secondary waterproofing membrane is therefore mandatory to keep storm water at bay. Today, this is usually an undertile membrane such as granule surfaced asphalt-coated felt with hot asphalt-sealed side and end laps. Tiles are adhered to the granule surfaced membrane with a sand–Portland cement mortar. Some tiles have fastener holes cast into the body of the tile, and some tiles have special metal hooks to improve their wind resistance.

The case in point was a Florida condominium. Pairs of units were roofed with relatively flat ceramic tiles set in mortar on an undertile membrane. The final ownership of the units, and care for the public spaces involved, had just recently been transferred to the Condominium Association from the developer.

Consistent with modern practice, the developer was sued for a long list of errors including, in this case, a report that the tile roofs were starting to slide off the building, endangering personnel as they entered or exited the units. (Suing the developer upon the transfer of the condominiums to the Condominium Association is almost a normal practice.) We were called in to investigate the sliding tile system claim.

We started on the roofs of several units, inspecting the tiles in place. On some units the applicator had shortened the exposure of the tile near the ridge, probably to avoid cutting a lot of tile. We saw no evidence of movement, such as a line of tiles sagging toward the eave. We removed tile in several areas. The mortar was firmly adhered to the undertile membrane and less substantially adhered to the underside of the tile, though no displacements were noted. We examined the edge of the undertile, where it overhung the metal edge flashing at the eaves by 6–8 mm ($\frac{1}{4}$–$\frac{1}{3}$ in.). The dent made in the bottom of the undertile by the metal edge flashing showed no displacement had taken place since the undertile was installed. We concluded that the alleged sliding never took place and that specific claim was without foundation.

QUESTIONS

1 Why do you think the sliding claim was made?
2 What remedial action is appropriate?
3 Short of not working for condominiums, what can be done to avoid this and similar problems?

31 The case of the skaters' cracks

Before the big ban on asbestos in the United States and the introduction of strong glass fiber felts, asbestos felts were widely used; particularly for smooth surfaced roofs. Asbestos felts were always slightly weaker in tensile strength than organic felts. As the awareness of the dangers of air borne asbestos increased, the Environmental Protection Agency clamped down on the asbestos in the effluent coming from the felt mills. The felt mills started using finer screens, and the resulting felt product contained a larger proportion of fine short fibers, further decreasing the strength of the asbestos felts.

We were called to investigate a roof on a single story manufacturing building in western New England. The steel framed building had a conventional bar joist and corrugated steel deck supporting a smooth surfaced asphalt-asbestos built-up roofing system. Above the steel deck, the bottom to top roofing consisted of:

- a 24 mm ($^{15}/_{16}$ in.) thick layer of glass fiber insulation board,
- a 19 mm (¾ in.) thick layer of perlite insulation board,
- an asphalt-coated asbestos felt base sheet,
- three asphalt-asbestos felt plies set in hot asphalt, and
- a glaze top coating of about 1 kg/m² (20 lb/square) of hot asphalt.

We saw stains on the floors, walls, and hung ceiling that was consistent with the general leakage reported. The owner reported that the current roof was installed the previous fall and that the roof was watertight until the early spring – as soon as the local springtime thaw was evident.

Some thin ice was still evident in bird baths (localized depressions) up on the roof. The roof had no noticeable slope. The roofing surface was drained by a gutter on the rear edge of the building. The edges of the felt plies were quite visible through the asphalt glaze coating and were about 290 mm (11 ⅓ in.) apart; consistent with the specification related to us.

The surface of the roof was covered with long curved cracks, each extending across several of the exposed felt plies. Part of the cracks extended only through the glaze coating; other cracks extended through at least the top ply felt. Some parts of the cracks extended through the membrane, and were the obvious source of the leakage reported. The crack pattern did not relate to any feature of the structure or the roofing system.

Our laboratory analyses of the samples removed during our field investigation revealed nothing that could explain the cracks we observed. To be sure, the asbestos felts used were some of the then new and weaker variety, but mere weakness does not explain the shape or the forces needed to cause the cracks.

After much study, we finally concluded that ice on the roof cracked in the pattern observed, and thermal cycling at the cracks telegraphed the cracks through the membrane.

QUESTIONS

1   Can you offer another explanation for the cracks?
2   What work was at fault?
3   How could have this problem been avoided?

32 The case of the phased felt plies

Phasing felt plies can be good and it can be bad. Phasing means to apply a layer of felts over the area to be roofed, and then apply additional felt plies to finish the membrane. Typically an asphalt-coated base felt and two or three ply felts, a two polymer modified asphalt roofing membrane, and a typical waterproofing membrane all have a phased construction. The difference between acceptable and unacceptable phasing is time and moisture. If the applicator applies all the plies during the same work day – no problem exists. Unacceptable phasing takes place when one or two plies are installed over the area on one day, and the remaining plies are installed later – even the following day, because overnight the surface of the first ply becomes coated with moisture. Rain isn't necessary; the morning dew is enough to create a blistering problem. Curiously, the morning dew on a roof usually signals a good day for roofing. Beware of rain if the membrane is dry in the morning.

Sometimes it is necessary to leave one or more plies exposed for a short time. In that case, a thin glaze coating of hot asphalt is recommended to seal the surface until the remaining plies can be installed.

The case in point is about a severely blistered roof on a discount department store in the middle of Massachusetts. We were called in by the owner because his new roof was covered with blisters and the roofing materials supplier recommended a fix that his thought was inappropriate.

The bottom-to-top components of the roofing system were:

- a single thick layer of glass fiberboard insulation,
- an asphalt-coated organic base sheet,
- three plies of asphalt-organic felt in hot asphalt, and
- an asphalt flood coating and gravel surfacing.

Many of the ceiling tiles inside the store were stained consistent with general widespread leakage. Plastic interior gutters channeled rainwater into barrels placed about the store. Up on the roof we saw many tubular blisters in the roofing membrane. The blisters followed the length of the felts and were about 860 mm (34 in.) on centers. Some of the blisters were quite hard (supported by the blister's internal pressure). Other blisters were soft; they had collapsed and many were open to water intrusion. We cut a number of samples from the roofing system for analysis in our laboratory. The blistering pattern did not match any structural feature of the building.

In the laboratory, by splitting frozen samples, we found the blisters were associated with lines of pin holed interply asphalt. The pin holes in the asphalt look like the craters on the moon; they are the residue of ruptured moisture bubbles. In this case, the pin holes lined up with the buried edge of the base ply – showing the base ply was damp or wet during the installation of the balance of the felts.

Two conditions are necessary for blistering. There must be a void or bubble to inflate and there must be a source of gas (almost always water vapor) to inflate the blister. I have seen both wet void free roofing membranes, and dry membranes without voids; neither had any blisters.

The roofing materials supplier offered to provide the material and labor to pressure-inject an asphalt mastic (asphalt, petroleum solvent, fibers and other fillers) into the blisters in return for an agreement absolving them from any future responsibility. The owner did not agree to this offer after he saw a demonstration of the injection process where the pump failed, and mastic oozed from every opening in the injected blister.

Large blisters can be repaired by cutting them open and filling the void within, with either mastic or hot asphalt. After careful repairs, about half of the blisters usually return, because it is almost impossible to fill all of the voids.

QUESTIONS

1 Whose work is at fault in this case?
2 Was the moisture found within the membrane due to phasing or due to wet or moist base sheets due to improper storage or shipment?
3 How can similar problems be avoided?

33 The cases of the fastener backouts

This is about two widely separated roofs with the same fundamental problem. The first was in Canada; the second in upstate New York.

We were called to investigate a roof on a manufacturing plant in Canada. The building was steel-framed with steel bar joists and deck supporting the roofing system. The roofing system consisted of several layers of insulation and a plate mechanically fastened to the deck and an EPDM membrane adhered to the plates. Leakage was reported everywhere inside the plant.

The large roof terminated at gravel stops and had no perceptible slope to interior drains. It had the usual fan curbs, sanitary vents, and HVAC equipment. The fasteners for the insulation were clearly visible through the membrane. Some fastener heads were down inside the plastic plates on the insulation. Other fastener heads and fastener heads with plates tented up the EPDM membrane. Some fastener heads had punched through the membrane, and were the obvious source of leakage. Visually, the larger the roof areas between penetrations and perimeters, the greater the number of displaced fasteners. Also, the degree of displacement appeared to be greater toward the center of the larger roof area.

We were called for the second job by the lawyer for the local district school. The stage in their new High School was leaking, and the lawyer was preparing to sue the materials supplier, the roofing contractor, and the designer for defective roofing system. We asked: "What investigations have been made, and what do the various parties have to say about the problem?" The lawyer responded that the designer knew the stage roof was leaking, but he had not gotten up there because no ladder was available; neither the roofing contractor nor the material supplier had been notified. We suggested that the roofing contractor assist us with our investigation – and provide the ladder needed to get to the roof.

The roofing system consisted of a thick layer of expanded polystyrene insulation, a polyethylene separator sheet, and an adhered plate 1.27 mm (50 mil) PVC (poly [vinyl chloride]) membrane, toggle-bolted through plates to a precast concrete deck. Toggle bolts tented the membrane upward in the center of the 560 m^2 (60 square) roof.

The toggles were missing from the toggle bolts (we later found them on the stage below). Individual bolts and plates were displaced upward, and the foam insulation was chopped up about the fastener hole.

In both of these cases, wind was the force that caused the fastener displacement. In the Canadian case, wind flutter caused the screws to rotate out of the deck. The rotation was permitted by the loose fit between the shaft of the screw and the plastic plates, and encouraged by the distance between the threads. The upward wind force on the screw brings the loose threads up against the deck and slides along the thread to rotate the screw. A repeated upward blow from the wind eventually frees the screw. In a similar fashion, wind flutter pummeled the toggle off the bolt. The wind lifted the bolt up out of the deck and insulation, to dance about the hole in the deck, and to mince the insulation.

As an aside, we settled the northern New York complaint by putting the toggles on the bolts backward (so the toggles were retained by the head of the bolt), inserting the toggle into the hole on the deck, pulling upward after the wings of the toggle expanded under the deck, bolting the plate and the insulation in place with common bolts, and cutting off the excess screw. The roofer patched the roof to finish the work and end the complaint.

QUESTIONS

1   Did the original design of these roofs properly take into account wind forces?
2   Was the fastener and plate design provided in Canada suitable for use?
3   What could have been done to prevent these problems?

34 The case of the blistered shingles

Blistered asphalt shingles are not new. The roofing industry has been plagued by blisters for many years. At various times blistering in shingles was thought to be due to:

- Low quantity of asphalt in the asphalt-organic felt core (usually expressed as a low percent saturation – 100 times the mass of the saturant divided by the mass of the dry felt).
- Low saturating efficiency – the percent saturation divided by the kerosene number – the kerosene number is said to be equivalent to the maximum mass of asphalt that can be absorbed by a specific felt.
- Excessive filler in the filled asphalt coating.
- Inadequate quantity of back filled asphalt coating.

Most asphalt organic shingles are made on 48–50 point felt. A 50 point felt is approximately 1.27 mm (50 mil) thick and weighs 50 lb/480 square feet – the felt gauge and also the dry felt ream. The percent saturation we normally viewed ranged from 150 to 230. Consider the following data in Table 34.1.

Table 34.1 Dry felt mass, percent saturation, and water absorbed

Dry felt mass		% saturation	Asphalt mass		Water mass at 90% RH		% water in felt
g/m^2	lb/100 ft^2		g/m^2	lb/100 ft^2	g/m^2	lb/100 ft^2	
488	10	120	585.6	12.0	39.04	0.8	3.64
488	10	140	683.2	14.0	39.0	0.8	3.3
488	10	160	780.8	16.0	39.0	0.8	3.1
488	10	180	878.4	18.0	39.0	0.8	2.9
488	10	200	976.0	20.0	39.0	0.8	2.7
488	10	220	1073.6	22.0	39.0	0.8	2.5
488	10	240	1171.2	24.0	39.0	0.8	2.4

Note
Calculated values.

Data of this sort led to the conclusion that the higher the percent saturation, the lower the percent moisture in the saturated felts – therefore the lower the blistering tendency. While completely accurate, the conclusion is false, because the motive blistering force is not the percent moisture in the saturated felt; it is the quantity of moisture present – which you note is constant.

Saturation efficiency is a more valid concept for retarding the absorption of water; the higher the saturation efficiency, the slower the absorption. But, our old nemesis water cannot be stopped. Currently, an acceptable saturation efficiency is 90 percent or greater for organic felts.

A concept was once offered that the higher the percent filler, the lower the blistering, because the stiffer and more highly filled coating would provide greater resistance to deformation. There may be a measure of truth here, but the higher filler loadings led to decreased durability through cracking.

Once it was fashionable to put most of the asphalt coating on the top of the shingle – where the weather resistance was needed. This "starved" the back of the shingles. Eventually, the industry realized that the backs of the shingles had to be sealed with coating to prolong the life of the shingle and to reduce the blistering tendency.

The increase in popularity of glass fiber based shingles, and the concurrent decrease in the organic felt based shingles, greatly reduced the incidence of blistering. At last! Blistering was a problem of our past because glass fiber based shingles never blister. Well, almost.

A roofing contractor friend called, and asked us to look at a blistering problem in glass fiber based shingles on a pavilion for a well-known theme park in Florida. His bill was not being paid because of the blistering. The pavilion designer's work was also in question, because of a missing vapor retarder, because everyone knew that glass fiber based shingles never blister, therefore someone else had to be at fault.

We examined the roof, and found the shingles were indeed blistered – mostly on roofs facing the south, but nothing else was out of the ordinary. We took blistered shingles off the roof and secured some shingles of the same batch that had never been installed from the roofer.

In our laboratory we ran all three of the blistering test available – only one test, a Canadian test involving prolonged water soak, and high temperatures in a vacuum oven, reproduced the blistering. Typical and detailed shingle analyses showed nothing out of the ordinary. We tried NMR (nuclear magnetic resonance) without usable results.

We finally tried digital fluoroscopy, or digital X-rays. Each film provided a plan view, and multiple vertical sections. The high density of the roofing gravel masked everything in the plan view, but the vertical sections were the prize. They showed small voids or bubbles on the glass felt in the unexposed shingles, and provided views in the exposed shingles of the inflation of the voids into blisters. It is probable that the felt dryer on the roofing machine, just prior to the coater, was inoperative, or the glass felts were unusually moist, or the binder on the glass felts was incompletely cured. In any event, the problem was not the pavilion designer or the roofing contractor.

QUESTION

1   What is the important lesson this case illustrates?

35 The case of "1+1=4"

Although it was mentioned briefly previously, the case of "1+1=4" is a perfect example of marketing hysteria. "1+1=4" was the banner heading of the advertisements by the originator of the idea that two asphalt-coated organic felt plies were equivalent to the then bench-mark standard four ply built-up roofing membrane. "The two coated plies have the same strength as the four uncoated ply felts," trumpeted the ads. Adding: "Although the coated felts cost more, the labor is less, so the owner is getting greater value, because more of the cost stays on the roof; it does not depart with the roofers." Aside from confusing cost with value, these claims were not true.

The asphalt-coated felt used 33–35 point felt. The dry felt reinforcement is 27 point for #15 asphalt-saturated organic felt. With a typical 140 percent saturation, a factory square of #15 felt (108 ft^2) weighs ~15 lb, hence the #15 name. (The 108 square foot factory square [36 in. × 36 ft] is an early unit of production that is still being used; it is enough material to cover 100 ft^2 of deck, with allowance for a 2 in. side lap and a 6 in. end lap.) Thus, the reinforcement in a typical four ply built-up roof is 108 felt points (4×27); much more than the 66–70 felt points used for these two ply roofs.

Far more important than the diminution of the membrane strength, was the fact that the two ply suppliers promoted phased construction. They told owners, general contractors, and roofers that they could quickly dry in a building with one ply of coated felt, and then later, after all the other work was completed, install the final ply, the flood coating and the gravel surfacing. Blisters and law suits abounded. But, that's not the worst.

The worst is the fear in the industry that the two ply introduction would cut into each competitor's market share. Couple that with the greed that recognized that the profits on a branded product under various trade names was much more than the profits possible from selling commodity felts, and a rash of "new products" – all two ply, quickly appeared. There were two plies based on 27 point felt, and even 50 point shingle felt. There were two ply early glass fiber felt, two ply asbestos felt, and even two ply coal-tar pitch systems rushed to market. Of all the suppliers, only two (one in the US and one in Canada) resisted the temptation.

The most economical-minded owners (read cheap) were quick to endorse and use these two ply systems with the lowest of the low bidders they could find. New and inexperienced roofing contractors were among the first to use these systems, and their inexperience and the quality of the workmen they used was evident in the final product.

The huge number of failures may have been the force for the development of the large number roofing consultants that appeared on the scene. Prior to the two ply syndrome, the number of competent roofing consultants in the United States could be counted on two hands. After the two ply syndrome, a huge number of people styled themselves as "roofing consultants" without much immediate improvement in competence. The RCI (Roofing Consultants Institute) was formed to improve this competence. Generally the variability in the competence of individuals who style themselves as roofing consultants is probably greater than the variability in the competence of lawyers or barristers.

The roof selected for this case is over a strip mall. The mall's owner complained of leakage and demanded replacement of the roofing system. We were asked to investigate the roofing system by the roofing materials supplier.

We found almost no evidence of leakage inside the mall. There were a few stained ceiling tiles.

Up on the roof, we found little of note. The existing roofing system was supported by bar joists and a steel deck; it consisted of two layers of insulation and a gravel surfaced two ply roofing membrane. The roofing surface sloped from the front to gutters at the rear of the mall. The roof was blister-free insofar as we could observe.

We cut six samples from this roofing system. In every instance, all components were dry to the touch and well adhered to each other and the deck. The deck was dry and free of rust. We abandoned sampling when new samples failed to reveal any new information. Our laboratory studies on the samples we retrieved showed all components were dry, were void-free, and the asphalt-coated ply felts were shingled (in shingled felts, the two felt plies are overlapped rather than being installed one ply at a time).

In deposition, the owner admitted that the former leaks were traced to alterations made by his tenants. He also knew several of his competitors had obtained new roofs for their two ply roofs "that everybody knew were unsuitable for the purpose."

QUESTIONS

1   What characteristics influenced the performance of this roof?
2   What do you think was the outcome in this case?
3   What point does this case make?

36 The case of the cold process roofing

Cold process roofing utilizes solvent-based adhesives and emulsions with reinforcing felts or fabrics to manufacture roofing and waterproofing membranes. It is an old technology that is hampered by economics. It obviously costs more to add solvents, fillers and fibers to asphalt to make roof coating and adhesives, or using high shear mixers to emulsify asphalt and water to make emulsions, than it does to sell the asphalt alone. This does not mean that cold process systems are not useful or valued. For example, the principal roofing systems for the Canadian military were all cold process. Cold process adhesives are currently being used to good effect to adhere the cap or top sheet in polymer modified roofing.

There are some firms that are an embarrassment to the roofing industry and have given cold process roofing a murky reputation. These include people who sell products with prices slightly higher than their extremely inflated claims. People who sell asphalt resaturants to renew the life of older roofs, and people who sell compounds that can take the place of whole roofing systems, are among the industry's detractors. You probably believe that eating oysters can improve your virility if you believe that any compound can renew the life of an old asphalt roof. The second group, the irresponsible sellers of "magic" roofing systems, is the subject of this case.

We were called in by the owner of a manufacturing building located in New England to investigate the roofing system that was currently being installed. He had purchased the system from a salesman who traditionally sold directly to the maintenance supervisors of industrial facilities. The owner was concerned because the roofing workmen on the job seemed to lack training and, more importantly, the new roof leaked.

The "specifications" listed on the purchase order required the removal of the existing roof, installation of a single layer of glass fiberboard insulation, taping the insulation joints, installing a combination glass fiber fabric and scrim in a cold process cement, and installing a mastic top coating and gravel surfacing.

Up on the roof, we interviewed the workman; we found the foreman, who was the chief spray man, had extensive experience (his words) in spraying damp proofing on foundations. This was his first roofing job. The other two workmen had less roofing experience.

Sampling was difficult because of the mastic top and bottom coating that had not completely cured. We managed to obtain several samples. In each case the insulation felt wet, and there was water in the rusty deck flutes.

In the laboratory we found the mastic attacked the adhesion of the tapes over the insulation joints – the tapes were not adhered to the insulation. The mastic was pinholed on the glass fabric, and the membrane was not watertight.

We made up samples in the laboratory using raw materials obtained on the roof, carefully measuring the quantities to insure compliance with the supplier's recommendations. The resulting samples were pinholed and not watertight.

QUESTIONS

1   What actions contributed to this disaster?
2   Who was at fault?
3   What remedial action is required?
4   How can this kind of problem be avoided?

37 The case of pressure sensitive adhesion

Pressure sensitive roofing products have been developed for many roofing applications including tapes, underlayment for shingles, waterproofing sheets, and lap seals for rubber products. They are attractive because they generally do not require special installation equipment, and sometimes obtain a better bond with the substrate than can be obtained by any other method.

Just one word of caution; do not rely on pressure sensitive materials unless they cure, harden, or are otherwise fastened in place. During application, high positive pressure for a short period of time is used to obtain adhesion. What is not usually realized is that a slight negative pressure for a longer time will destroy the adhesion. The product of the time and the temperature is probably a constant for a true pressure sensitive adhesive.

Here are two cases to illustrate the point. The first is a roofing system developed by a major supplier of roofing products. They developed a single ply product consisting of a highly durable plastic sheet bonded to asbestos felt. They also contacted a major pressure sensitive tape manufacturer, who developed a tape exclusively for use in the roofing system.

The initial installations were both beautiful and effective. The tapes over the joints started falling off after only a short exposure, and the great contrast between the original appearance and the appearance after the tapes fell off guaranteed complaints. We found that silicone sealant could be used to adhere the tapes, but any silicone out beyond the tapes made the local dirt water repellent, and the resulting dirty streaks were esthetically as bad as the missing tapes. The expense of the silicone sealant was too high for the market. Eventually, this theoretically valuable roofing system had to be withdrawn from the market.

The second example involves the misuse of materials. We were asked to investigate the roof on a new Junior High School, deep in one of the Southern States. We were told the roof involved a steel deck, lightweight insulating concrete, and a base and two asphalt-asbestos ply felt built-up roofing. With this background we felt we would find huge number of leakage sources – a true heaven for a roofing investigator.

We were totally wrong. The existing roofing system drained nicely, the steel deck was slotted to drain the lightweight insulating concrete, and there were no blisters or splits in sight. We finally traced the leakage to the perimeter parapet wall flashing. The wall flashing consisted of a single plastic faced – polymer modified asphalt – self adhering membrane intended for use as water proofing, shingle underlayment, or window flashing. Thermal cycling had opened all the side laps and was in the process of moving the sheets off the wall. We never learned if the flashing was designed the way it was installed, or if the installation was some mental aberration of the installer.

Pressure sensitive systems can do valuable work as concealed window, door, and other wall penetration flashing, as concealed waterproofing, and shingle underlayment. Failure will result with prolonged exposure to the weather.

QUESTIONS

1   What lessons do these cases provide?
2   How could have these problems been avoided?

38 The case of the fire retardant plywood

Fire retardant plywood is not a roofing product per se, but it is frequently used in conjunction with roofing products. Building Codes as an example, frequently require fire retardant plywood at party walls between condominiums or apartments, and on the roof deck about 2 m (6 ft) on each side of the party wall. Fire retardant plywood works by early charring. Charred wood retards flames. Salts impregnated into the wood, triggered by heat, cause the early charring.

Problems arise when economically minded builders or uneducated occupants vent bathroom fans, clothes driers, or kitchen fans into attics instead of outside. Heat built up from these sources can, like a fire, trigger early charring in fire retardant plywood. The charring is accompanied by a significant loss of strength, permitting loads (like people) to fall through the deck. In some relatively rare isolated cases of defectively manufactured plywood, early charring occurs even without the improper venting.

We were called in on what was reported to be a fire retardant plywood deck problem at a condominium complex in the Middle Atlantic States. This series of connected 10-year-old town house style condominiums were wood framed, with the usual fire retardant plywood provisions at the party walls, and roofed with asphalt-glass fiber shingles.

Our field investigation showed that the seal tab feature of the shingles were firmly adhered, and many of the shingle tabs were split (see Chapter 25). The plywood deck near the party walls was in poor condition; it was partly collapsed between the joists in some areas. Attic inspection showed typical soffit and ridge vents, and spaces where the owners had vented bathrooms into the attics. Attics with and without bathroom vents showed the same degree of plywood deterioration. There were the same number of splits in the roofing over attics with and without bathroom vents and in roofing over regular and fire retardant plywood.

Laboratory tests of the samples recovered from the roofs showed the shingles were quite brittle and had very low (~500 gm) tear strengths.

The *roofing supplier* blamed the thermal splits on the improper venting of the attic spaces; he pointed out that the warranty specifically excluded coverage when the deck was found to be improperly vented or found to be unstable.

The *fire retardant plywood supplier* said that the plywood problems were due to the bathroom vents pumping hot, moist air into the attics. If they were found liable for the fire retardant plywood, they might pay for labor and new plywood, but they were unwilling to pay for the removal and replacement of the defective shingles.

The *Condominium Association* demanded replacement of all the shingles (partial replacement would result in a patchy appearance and, in any event, all the shingles had failed) and all of the fire retardant plywood at no cost to the condominium owners.

QUESTIONS

1   Who should pay for what?
2   What are the root causes of these problems?
3   How could these problems be avoided?

39 The case of asphalt dispersion

Polymer modified asphalt (PMA) is made by dispersing soft asphalt in an APP, SBES, SBS or similar polymer. The polymer is mixed into the hot asphalt where it swells and a phase inversion takes place. The original mixture is particles of polymer in a sea of asphalt (Figure 39.1). After inversion, the mixture becomes particles of asphalt in a sea of polymer (Figure 39.2). Additional mixing prepares the final fine dispersion (Figure 39.3) similar to a mayonnaise. Properly modified asphalt has enhanced strength and elasticity, and a much higher softening point than the starting asphalt. The original asphalt and polymer selection, material proportions and degree of mixing are critical for the stability of the final product. Improper materials, proportion, or mixing may result in a mixture without phase inversion, or an unstable product that will revert toward the original components. Good dispersion does not guarantee good performance, but less than ideal performance can be expected with poor dispersion.

We have proposed a dispersion standard in ASTM International, but thus far our test method has not been accepted.

We were approached by a Federal Agency that was having a problem with a polymer modified asphalt roof. Foot traffic was leaving footprints in the soft granule surfaced membrane. We obtained samples of both the exposed and the unexposed granule surfaced sheets for laboratory analyses.

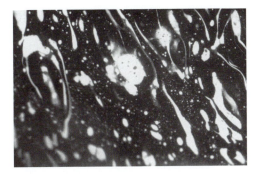

Figure 39.1 Polymer in a sea of asphalt.

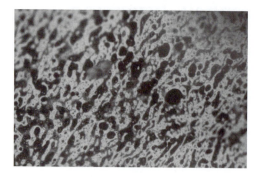

Figure 39.2 Just after phase inversion.

Figure 39.3 Final dispersion of asphalt in a polymer.

Our laboratory data showed the composition of the membrane was appropriate, but our check of the asphalt dispersion with a microscope and an ultraviolet source showed the dispersion was poor at best.

QUESTION

1   How could have this problem been avoided?

40 The case of the liquid applied waterproofing

The owner of a very large building in a major city on the Eastern Seaboard asked us to look into a plaza-waterproofing problem at his building. He wanted to sell the building, but was unable to do so because of the leakage.

The original plans called for a generous plaza including a fountain, a waterfall, a meandering stream, and a row of large trees set in planters. All this is to be over a buried parking garage. The specifications called for a liquid applied coal-tar extended polysulfide-waterproofing system.

We went to the site during the summer and found the fountain, stream, and waterfall turned off to minimize the leakage into the garage. All the trees in the planters were dead or dying. A cooperating masonry contractor opened the deck. We were amazed to find this bottom-to-top construction:

- The structural concrete slab.
- A relatively thin trowel coating of the coal-tar polysulfide compound without any reinforcing. The cured membrane was adhered in some areas, and unadhered, blistered, and pinholed in other areas. Some of the unadhered surface of the membrane had the lumpy appearance usually associated with the surface of a human liver or brain.
- An unadhered layer of polyisocyanurate board insulation.
- Thin pavers grouted in over the insulation board.

When we opened the plaza at a drain, we found water pooled everywhere. There was no evidence that drainage was provided at the waterproofing membrane level; the only drainage was at the surface of the plaza. We inspected the plans and found that the construction was as specified. We checked the drawings that involved the large planters at the perimeter of the plaza. Planter drains were not specified.

Major leakage was reported every time the fountain was put into operation. We therefore carefully inspected the features of the fountain that were visible and found no sign of the large hole required to provide the reported leakage. With much discussion, we persuaded the contractor to dissemble the fountain's water intake because we could not see how the intake flange was fastened to the plaza's structure. The stainless steel intake was about

380 mm (15 in.) in diameter pipe welded to a hole in a 600×600 mm (2×2 ft) plate, about 19 mm (¾ in.) thick. The square flange had no fasteners to the deck. We saw no evidence of flashing or waterproofing under the flange. The hole in the concrete deck appeared to have been jack hammered out, leaving rough concrete without any waterproofing.

We persuaded a friendly plumber to remove several of the plaza drains to permit the accumulated water drain away.

QUESTIONS

1   What actions should be taken, considering the existing conditions?
2   Was the design defective? How?
3   Was the workmanship defective?

41 The case of the blistered airport roof

Blisters are likely to be found in roofing systems composed of any material; no system can be said to be blister free. We have seen blisters in built-up roofing, asphalt-organic shingles, asphalt-glass fiber shingles, and even 3 mm (¼ in.) thick sheet lead.

Remembering our previous lessons, blisters can only occur in systems when there is a void or pocket to inflate and a source of gas (most often water vapor) to do the inflating. Our friendly star, the Sun, provides the power. We were therefore taken aback when we were asked to investigate a blistered single ply roof on a Southern airport.

Thermal insulation for air conditioning was installed under the roof. The roof deck was poured-in-place high strength concrete. After priming, the one ply roof was fully adhered to the deck. Blisters started to appear shortly after the membrane was installed. Some blisters were quite small – about 45 mm (1¾ in.) in diameter; other blisters were much larger, up to about 600 mm (2 ft) in diameter. The blisters were randomly distributed (meaning that we had no idea why they appeared at their locations).

The usual roofing membrane samples told us nothing about the blistering. We found primer and dried adhesive under every sample. The concrete deck was smooth and unblemished.

We finally resorted to a coring machine to obtain samples of the membrane, adhesive, primer, and concrete for analysis in our laboratory.

Vertical sections through the concrete cores revealed small pockets on the surface of the concrete. These pockets appear as small holes on the surface and enlarge into bottle-shaped voids just below the surface. The necks of some of the bottle-shaped voids were lined with dried primer.

At last we had an explanation for the blistering. The concrete deck was beautifully finished, probably with a steel trowel, leaving very small pock marks on the surface. Primer application covered, but did not fill the pock marks. The wet primer in each hole was probably in a double concave vertical section – the center of the concavity thinning and finally rupturing as the primer dried. The much more viscous adhesive and the flat bottom surface of the single ply bridged and sealed the void. Solar heat expanded the gases in the void consisting of air, moisture, and perhaps some of the adhesive's solvent vapors, to expand the blister. Nocturnal temperatures caused the trapped gases to lose volume – probably creating a partial vacuum to draw in additional moisture from the concrete surrounding the void – to reinflate and expand the blister in the next solar cycle. Removing the existing membrane and installing a fully adhered polyester fleece-backed membrane solved the problem. Fleece backing relieved local pressures that caused the blistering.

QUESTIONS

1   What lesson does this case teach?
2   How can you avoid this problem in the future?

42 The case of pesky sea gulls

Sea gulls can be a pest in almost any marine environment. Up on the roof they can eventually cover the entire surface with their debris and their bodies.

We were called out to investigate a roof along the New England sea board; not for the sea gulls, but for the leaks in the roof. The roof was on a single story principal store of a strip shopping mall located about halfway between the town dump and a beach on the ocean. The roof made a perfect landing craft for sea gulls. The bird baths on the dead flat surface I'm sure were appreciated by the gulls.

We traced relatively few leaks directly to the birds. We found that the birds had pecked holes in some of the expansion joint covers and a few of the flashing systems at penetrations. They seem to be attracted by white materials peeking through the black EPDM.

The majority of the leakage was through inadequately adhered laps in the EPDM. Early in EPDM history, a neoprene-phenolic contact adhesive was used. Normal application at the time was to:

- solvent clean the surfaces to be adhered,
- scrub a thin coating of adhesive on both surfaces,
- allow the adhesives to become tacky, and
- put the tacky surfaces together and roll the seam with a roller to knit the joint.

Many technicians believed that a thinner adhesive layer developed a stronger bond, much as a thin layer of water between two plates of glass provides a very strong attachment. Testing cured sample seams with increased adhesive applications proved the theory wrong. The thicker experimenters made the adhesive layer, the stronger the bond between the rubber sheets.

Generally, the splicing technique described, worked quite well on most roofs. We monitored the roofing work on many roofs during this interval, and none failed. As time advanced, some suppliers believed that it was only necessary to apply the adhesive to only one rather than both surfaces. This roof was the result.

> *Big cost improvements*
> *Depreciate quality*
> *Until nothing works.*

The owner tried an interesting experiment to rid himself of the sea gulls. After we had the old roof replaced, the owner gathered a large number of plastic $0.02 \, \text{m}^3$ (5 gal.) pails, cut drainage holes in the bottoms, weighted each pail with several large rocks, and installed the pails on the roof at the corners of a 7.6 m (25 ft) square grid. He used the bail on each pail to string his grid with common iron wire. The sea gulls do not land on the roof anymore; I'm told they need a given distance to take off, and they are not at all happy with the wire grid.

QUESTIONS

1   How could the leakage be avoided on future work?

43 The case of the distant expansion joints

A roofing contractor friend called us out to investigate a puzzling series of roofing failures involving his best client's roughly 18,000 m² (~2000 square) roof.

The entire building is steel framed with bar joists and a steel deck supporting a roofing system with the following bottom-to-top construction:

- A mechanically fastened layer of composite insulation. The insulation is composed of 70 mm (2¾ in.) thick isocyanurate foam and 13 mm (½ in.) thick perlite/wood fiber in square panels 1.22 m (4 ft) on a side.
- A four ply ASTM Type IV asphalt-glass fiber felt shingled membrane built up with steep asphalt, and with a hot asphalt glaze coating.
- An aluminum pigmented asphalt roof coating was added sometime later.

The entire building was built in segments over several years. A ~6000 m² (653 square) building was built the first year. The building was 62 m (204 ft) north–south, and 98 m (320 ft) east–west. The roof sloped gently from a central ridge toward the east and west; it had no expansion joints. The western half contained 12 skylight – smoke vents evenly distributed over the roof in addition to some HVAC equipment. The eastern half of the roof has almost no penetrations.

A 7000 m² (742 square) addition was made the next year. This 71 m (232 ft)×98 m (320 ft) addition contained 25 skylights distributed over the surface; an expansion joint cover was at the intersection between the new and the former year's work. The same roofing system as used on the original work, applied by the same roofing company, was installed.

The following year (year 3) a ~5000 m² (560 square) addition was added. It had no skylights and very few other penetrations. It was separated from the former year's work with an expansion joint. Again, the same roofing company installed the same roofing system.

In years 4 and 5 leakages through splits were reported in the eastern slope and on the ridge of the first year's roofing work. All splits were over insulation joints. There were no splits in the areas with skylights.

In year 6 leakages were reported through splits in the four-year-old third roof area, and the roofing contractor called us in to determine the cause of the membrane splitting. We visited the site and observed the removal of 15 roofing samples.

The roofing membrane felt plies were applied in a north–south orientation. The surface of the roof was coated with aluminum pigmented asphalt coating in an effort to prevent cracking. We saw north–south patches on both sides and parallel to the ridge. There were five long patches on the west slope and three on the east slope. The distance from the center of each patch to the ridge was a multiple of 1.22 m (4 ft). We found gaps of up to 25 mm (1 in.) between insulation panels directly under the splits. The holes made by the fastener in the insulation were oval, with the major dimension of the oval perpendicular to the ridge. The aluminum pigmented top coating was striated parallel to the splits.

In the laboratory, the membrane analyses were not remarkable; they were dry and void free. We prepared laboratory built-up membrane samples using ASTM Standard D2178 Type IV and Type VI felts and heat conditioned half the prepared samples to simulate weathering. We measured the load strain properties of ten of the samples at −18 °C (0 °F), together with the laboratory prepared samples. Table 43.1 shows our data. These data show the ~four-year-old field samples are significantly weaker than the laboratory prepared, and the laboratory prepared and heat conditioned samples. There is no doubt that roofing system mobility down slope in response to solar cycles caused the splitting. This mobility overcame the combined shear resistance provided by the system's fasteners. This mobility could have been minimized by increased fixity as shown by the crack free areas with skylights; these areas never presented a problem. Additional north–south expansion joints might help, but only with provision for interior

Table 43.1 Load–strain testing of built-up roofing samples

Sample	Mean load at break		Elongation (%)
	kN/m	lb/in.	
Field samples			
Mean	33.3	190	1.9
Estimated standard deviation	5.7	32.3	0.45
Laboratory prepared			
Type IV felt, as prepared	56.7	324	4.2
Type VI felt, as prepared	81.7	467	4.8
Type IV felt, heat conditioned	54.4	311	3.4
Type VI felt, heat conditioned	71.2	407	4.0

drains to handle the drainage the new expansion joint would block. The weakness of the membrane and stress concentration over the moving insulation joints contributed to, but did not cause the cracking.

QUESTIONS

1   Who was at fault?
2   How can this problem be avoided in the future?

44 The case of the ill wind

After water and sun, wind is perhaps the greatest exposure enemy of roofing systems. Unless the roofing system is carefully designed and competently installed, wind can raise havoc.

A materials supplier called us in to investigate a wind loss problem at a mid-continental airport. The contract documents required, in part:

- Two layers of insulation mechanically fastened to a steel deck. The top insulation layer is to be fiberboard, and all insulation joints are to be staggered.
- A water-based adhesive is to be applied to the insulation and to the single ply membrane. The membrane is to be pressed into place after the adhesive becomes tacky. A solvent-based adhesive is to be used in colder weather.
- The roofing system installed shall pass the "Negative Pressure Tests," each day as described by the Factory Mutual System Loss Prevention Sheet 1–52.

The job records showed that the system complied with an I-90 rating during each daily test, but we found the membrane attached only at the perimeter of the roof and at penetrations; it ballooned upward everywhere. The local weather reports spoke of maximum wind gusts of 72 km/h (45 mph) during the preceding period.

We removed a number of samples of the membrane, insulation, and adhesives for our laboratory studies. We examined the membrane and insulation samples in the field. The contact surfaces both had a glaze, consistent with adhesive application. The glazed areas felt dry and non-tacky.

In the laboratory we measured the quantity of adhesive on the underside of the membrane by weighing a specifically sized sample, erasing the adhesive, and weighing the cleaned sample. The before erasure mass less than the after erasure mass provided a good estimate of the dry adhesive present per unit area. Weight per gallon and mass percent solids data enabled us to calculate the quantity of adhesive applied to the sheet. These data were quite consistent; they showed only half of the adhesive specified was applied.

To determine the adhesive in the insulation we weighed pieces of insulation from the field (with an adhesive application) and pieces of adhesive-free insulation. These data showed that the specified quantity of adhesive had been applied.

We next prepared laboratory samples, using the specified adhesive quantities on the contact surfaces. The adhesive penetrated into the insulation quickly and dried to a non-tacky surface. The under surface of the membrane adhesive was tacky. We obtained a firm bond by mating these surfaces.

We also prepared laboratory samples using the quantities of adhesive we had found in the field samples. We obtained a slight bond by rolling these surfaces together. We tried coating the insulation, and coating it again after the first coating had dried. We obtained a very firm bond between the coated and dried membrane and the double-coated insulation. Much of the adhesive applied sank into the insulation, unless the insulation was first coated to prevent penetration of the second coating. A bond could be achieved with the specified adhesive quantities, but adhesion was very sensitive to variation in adhesive applications.

QUESTIONS

1   Who was at fault?
2   What should be done to prevent this type of problem?

45 The case of the improper waterproofing

The irate maintenance manager for a city hall asked us to investigate the plaza waterproofing over the buried parking garage for the city hall. Car finishes were being ruined by the alkali water draining onto the parked cars. The cars were owned by unimportant people like the mayor and the various committee heads.

An interior investigation confirmed the reported leakage. We saw water dripping off stalactites dangling from cracks in the precast concrete deck.

Some of the pavers on the plaza in normal walking areas were in poor condition consistent with salt and freeze-thaw attack. From top-to-bottom we saw:

- The pavers.
- A mortar setting bed.
- A polyethylene film.
- A thick layer of hydrated bentonite jell.
- The concrete deck.

Everything below the pavers was soaking wet.

Bentonite is a clay (sodium montmorilonite) that expands many times with the addition of water. A four mass percent mixture pours with great difficulty. An eight percent mixture does not pour. To be effective as a waterproofing agent, the bentonite must be confined under pressure such as $1.4 \, kPa$ ($30 \, lb/ft^2$) and cannot be subjected to alternate wetting and drying. Sodium chloride inhibits the swelling of the clay.

We prepared a model of the plaza waterproofing system in our laboratory and subjected it to cycles of heating and drying. We periodically tested for leakage by following the path of water containing a small amount of methylene blue dye. Cracks in the bentonite quickly became apparent.

The *bentonite supplier* claimed that the waterproofing failed because of the deicing salts used on the plaza inhibited the swelling of the bentonite. He also said that the mass of the materials above the bentonite was not enough to properly confine the clay.

The *owner* responded that the supplier had ample time to examine the plans and specifications. The supplier should have objected before the work was started if the mass of the materials over the bentonite was insufficient. The bentonite was pre-swollen when it was applied, so deicing salts could have little effect on the clay.

QUESTIONS

1 Which work was at fault?
2 How could have this problem been avoided?
3 How can we prevent this problem from happening in the future?

46 The case of the poorly vented roof

All of the shingle manufacturers in the United States exclude roofs over poorly vented attics from their warranty. The concept is that the increased heat due to the lack of venting will materially shorten the life of the shingles, inducing problems such as thermal splitting of asphalt-glass fiber shingles. Lawyers tell us that impractical or unreasonable contract or warranty provisions may not be supported by the court. We therefore decided to investigate the relative validity of the suppliers' position.

Aside from the degree of venting, other factors that can influence the temperature of shingles are the geographic location of the building, the color of the roof, the orientation of the exposure, and the slope of the roof. We calculated the temperature of the shingles in seven locations (Green Bay, Wisconsin through Miami, Florida), for white and black shingles, for five orientations (90° through 240° – east through south to west), for 25–100 percent slopes (3–12 in./ft), and three degrees of ventilation. The three degrees of ventilation are:

- No venting provided.
- Venting equivalent to 1/300 of the plan area (a rather arbitrary value generally accepted as appropriate venting).
- Venting equivalent to 1/300 of the plan area assisted by a wind perpendicular to the eave.

We calculated the temperature difference of shingles on non-vented and vented constructions as 0.44–0.61 °C (0.79–1.1 °F), and the temperature difference between non-vented and vented with a wind assist constructions as 0.69–0.85 °C (1.24–1.53 °F).

We calculated the temperature difference of black and white shingles with each type of venting. The difference between the temperatures of black and white shingles on an unvented deck is 1.22–1.67 °C (2.2–3.01 °F), on a vented deck is 1.11–1.51 °C (2–2.74 °F), and on a vented deck with a wind assist it is 1.07–1.49 °C (1.92–2.69 °F).

The difference in the temperature between the shingles facing south and those facing west is 1.06–1.74 °C (1.9–3.14 °F) on an unvented deck,

0.96–1.58 °C (1.73–2.84 °F) on a vented deck, and 0.93–1.56 °C (1.68–2.81 °F) on vented decks with a wind assist. Shingles are minimal, at best.

The maximum difference in the temperature of exposed at 25–100 percent slopes is 0.14–0.39 °C (0.25–0.7 °F) for unvented decks, 0.09–0.55 °C (0.15–0.98 °F) for vented decks, and 0.15–0.31 °C (0.27–0.55 °F) for wind assisted vented decks.

Of the five parameters studied, the venting area had next to the last position of importance in controlling the temperature of the shingles. Venting had about 1/3 the influence of the aspect or color and 1/36 the influence of the geographic location. Even with the assist from the wind, venting reduces the average temperature of the roofing only half as much as changing from black to white shingles. Based on these data, the warranty exclusion for improper venting appears to be without merit.

Venting is important to remove any excess moisture from the space below the deck, but that moisture should have been excluded by a functioning air and moisture barrier. Be careful; in southern states venting can add moisture to the system rather than to remove it. We have many cases from which to choose, because the "poor venting" excuse rears its head during almost every case.

The School Board called us in to investigate the shingle splitting observed on the Junior High School roof in a New England state. The supplier invoked the poor venting excuse and refused to take any effective action.

We found split shingles over both vented and unvented decks. The tear strength of these four-year-old shingles was down to 450–550 gm (ASTM D3462 requires a 7000 gm minimum tear strength for new shingles). The supplier finally agreed to replace the shingles with shingles of a similar quality. I am told that these three- and a half-year-old shingles are now starting to split.

QUESTIONS

1   Is the poor venting excuse a proper warranty exclusion?
2   Can you see the danger in settling for replacement with like quality?
3   How can we avoid this type of problem?

47 The case of the missing facer adhesion

Of the various thermal insulation boards used in roofing, we recommend a cover board or second layer when glass fiber, polyisocyanurate, extruded polystyrene, and expanded polystyrene insulations are used for the first layer. This concept does not meet with universal approval – many suppliers believe their foam product can go it alone, without a cover board, but they are wrong for the following reasons:

- The denser and stronger cover board armors the membrane against impacts. The more impact resistant the system, the more likely it is to survive both the intentional and unintentional abuses during construction and service. As an example, when built-up roofing systems are exposed to severe hail, the membrane applied on glass fiber insulation will probably be punctured, while the same membrane over cover board and glass fiber insulation will survive.
- Insurance recommendations recommend mechanical attachment for all systems on steel decks. When the first insulation layer is mechanically fastened, adhering a cover board over the fastened layers before adhering the membrane separates the fasteners from the membrane and prevents fastener back out. As a general rule, try to minimize the contact between mechanical fasteners and the membrane to minimize points of stress concentration.
- Two or more insulation layers with staggered joints are recommended in all applications to prevent thermal breaks and stress concentrations where joints are aligned. Make the top layer a cover board.
- A cover board is always necessary when using polyisocyanurate foam insulation, because of the low cohesive strength of the foam. The tee peel strength of the facer off even good foam is in the range of 175 N/m (1 lbf/in.) width. It is lower for lesser quality foams. The slightest traffic tends to rupture the foam cells directly under the top insulation facer, creating a void that frequently blisters the membrane and facer by internal pressure, or by wind flutter.

With all these benefits, why do some people still want to use only one layer of insulation? The most probable answer is misguided cost savings, ignorance, or greed. None of these seem satisfactory. Let us look at the case in this chapter.

A new large factory was constructed for a major computer manufacturer who retained us to review the plans and specifications and to monitor the roofing work. The contract documents called for a vapor retarder, a lower layer of polyisocyanurate foam insulation, a cover board, and a fully adhered EPDM membrane. So far, so good.

The successful (low bidder) roofing contractor offered a relatively minor price reduction to eliminate the cover insulation board – slightly increasing the thickness of the foam insulation. We objected to this cost improvement, but the owner gave his approval after consulting with the materials supplier. The material supplier approved the cost improvement. Note that clients do not always agree with their consultants – particularly when money is involved. There is that endless conflict between "better" and "less expensive."

We noticed and reported blisters in the roofing after about half of the 56,000 m² (6000 squares) of roofing were completed. The rest of the roof had a cover board, and then the first half was replaced.

QUESTIONS

1   Who is at fault?
2   How could we have been more persuasive?

48 The case of the noisy roof

We were called in on not one, but two noisy roofs. The developer reported loud noises from each roof at dawn and sunset, and even when clouds came across the sun to shade the roof. Noises were not heard when the roof was covered by a thin layer of snow.

Both roofs had fully adhered EPDM membranes over a mechanically attached single layer of polyisocyanurate foam insulation. One roof had a 0.25 mm (22 gage) steel deck and 100 mm (4 in.) thick insulation mechanically fastened to the steel deck with approximately three fasteners persquare metre (one per four square feet). The second roof had a 0.9 mm (20 gage) steel deck and 127 mm (3 in.) thick insulation mechanically fastened to the steel deck with the same number of fasteners as the first roof, but with a different fastener pattern.

We know that noise is always the result of movement. Now the question is: What movement? If the noise were due to crushing insulation, or another similar destructive event, it would have died out as soon as the destruction was completed – but the noise continues. This argues for tin canning of the deck, or a similar repetitive type of event being the cause of the noise.

We contacted the insulation supplier and found that they recommended against installing thick insulation in a single layer. The supplier was not very forthcoming with other instances of noisy insulation, but we gathered that some similar events have occurred elsewhere. The supplier offered to coat the black EPDM roof white, with an acrylic coating in return for a variant on a hold harmless type of agreement. The owner has not agreed to that suggestion.

Site visits confirmed the noisemakers' presence. We cut samples for analysis and instrumented the steel deck to measure any movements present. The fastener holes in the insulation boards were ovated. Accelerometer measurements were small to non-existent even when the noises were quite loud. The temperature of the EPDM was measured to vary 61 °C (110 °F) between day and night during three days in May.

In the laboratory we measured the flexural strength of the insulation to failure. The paper facer of the insulation ruptured suddenly with a loud cracking noise that was quite dissimilar to the noises heard from the roof. In another test, we pressed the insulation against a steel surface and moved it slightly, and thereby generated a noise similar to the noise bothering the tenants of the buildings. The noise was louder as we increased the pressure on the insulation.

We concluded that the offending noises were caused by the desire of the insulation to warp upward in response to solar radiation. Since the panel was held flat by the fasteners, the bottom of the panel was forced to slide across the deck – creating the noise.

QUESTIONS

1   How can this problem be repaired?
2   What work is at fault?
3   What do you think about the suppliers' remedial offer?
4   How can this problem be avoided in the future?

49 The case of the severe hail storm

The annual losses due to hail have increased dramatically. In the last 25 years sensational losses have been posted in Australia, South Africa, North America and Europe. A 1976 Sidney hailstorm loss of 40 million dollars was regarded as a unique catastrophic event, but three years later a 20 million dollar storm struck Adelaide. Additional major storms include:

- 1980, South Africa, $10 million,
- 1982, United States, $200 million,
- 1982, Canada, $100 million,
- 1983, South Africa, $15 million,
- 1984, South Africa, $30 million,
- 1984, Germany, $500–$1000 million.

Early on in the US, most of the losses were due to destroyed crops, but now people and manufacturing plants have moved into the sunbelt areas that are prone to hailstorms. The claims to the insurance companies have increased enough so that steps are being taken to identify roofing systems that are hail resistant, and the insurance companies are tending to investigate claims that they previously paid without investigation.

Hail always occurs in thunderstorms, but not all thunderstorms have hail. Super cooled droplets dance in the wind and plate on layers of ice to form hail. Hailstones are roughly spherical in shape and vary in density. They fall from the sky when their mass exceeds the wind pressure holding them up. If you know the diameter of the hail, its impact can be estimated with reasonable accuracy since it falls at critical velocity. A 10 mm (⅜ in.) diameter hailstone has a terminal velocity of 50 km/h (31 mph), a 50 mm (2 in.) diameter hailstone has a terminal velocity of 110 km/h (68 mph), and a 140 mm (5½ in.) diameter hailstone has a terminal velocity of 170 km/h (106 mph). Generally hail of less than 32 mm (1¼ in.) in diameter, with an impact of 5.4 J (4 ft lb) will not do much damage to most roofing. On the other hand, 51 mm (2 in.) diameter hail, with an impact of 30 J (22 ft lb) will do damage (Koontz 1991). The threshold for damage seems to be about 20 J (15 ft lb). Fortunately large hail falls infrequently. A three-year study in Alberta, Canada showed that stones with 30 mm (1.2 in.) diameter were recorded 1500 times less frequently than stones with a diameter of 10 mm (⅜ in.).

An insurance company called us to investigate a hail damage claim made on behalf of three schools in Virginia. The claim was for new roofs on all three schools plus glass replacement for the glass on a rooftop greenhouse. The claim was accompanied by an engineering report commissioned by the school system. The engineer's audit was conducted eight months after the alleged hailstorm.

He reported the *High School* roof was supported by an intermediate rib steel deck, and consisted of a plastic vapor retarder, 50 mm (2 in.) of perlite roof insulation, and a four ply asphalt-organic felt roofing membrane with a gravel surface. "Portions of the roof have repaired by what appears to be an application of solvent reduced asphalt applied in layers with reinforcing mesh and surfaced with aluminum coating."

The report continued to report that school personnel observed 76 mm (3 in.) diameter indentations, 25 mm (1 in.) deep in the roofing. After a period of time the roofing returned to a flat condition. (This implies a hardly believable 83 mm (3¼ in.) diameter hailstone, and an improbable recovery.)

The damage to the *Elementary School* was similar to the damage to the High School roof; the damage to the *Junior High School* was not as severe as the other two, but flashing damage was evident.

Our field investigation revealed that the roof on the *Elementary School* had been damaged by hail, but that it had been in poor, blistered condition prior to the storm. The custodian said that the roof had been leaking for years (confirmed by the number of old patches observed on the surface), that no new patches had been added since the hailstorm, and that no new leaks had been observed.

We saw patched curb flashing on the upper roof on the *High School*. The glass in the small green house had not yet been replaced; from the deteriorated condition of the green house, there is a question whether it had recently been in use. Dented fan cowlings confirmed hail exposure, but none were damaged enough to require replacement.

We found almost no signs of hail on the *Junior High School* roof. A thin gauge aluminum wrapped pipe was slightly dented, and did not have any holes. Of 15 fan and man hatch curbs examined, only two showed minor hail damage.

QUESTIONS

1 What compensation or reward, if any, is due to the School Board for each school?

2 What action is needed in areas prone to hail damage?

50 What have we learned?

A review of the previous 40 chapters that present cases shows there are specific repetitive elements; actions that might have been taken which might probably have prevented the failure from taking place, or at least minimized the event's impact. In descending frequency of occurrence, here are the principal repetitive elements:

- Have an effective peer review,
- Use systems that have a successful track record,
- Use monitors hired by the owner to oversee the installation,
- Buy competence in your supplier, designer, and contractor.

Peer review means having the contract documents reviewed before bidding and reviewing any alterations made during or after the bidding process. Perhaps it also means listening to the reviewer and not ignoring advice. One third of the problems presented could have been prevented with peer review conducted by a dedicated and experienced reviewer.

An appropriate peer reviewer should be a specialist in roofing technology; he or she is likely to be a designer of roofing systems, with a long, successful track record in designing roofs that do not leak. Probably, but not necessarily, your ideal peer reviewer is a professional engineer or architect (most architects and engineers do not qualify by experience, training or interest). A nit-picker rather than a generalist is preferred, to be sure that every aspect of the work is properly detailed and integrated. The appropriate peer reviewer has no financial interest in any of the parties involved in the work in any way.

The reviewer does not take the place of the designer of record in assuming the responsibility for the ultimate design – this must remain with the designer of record. The reviewer rather augments the designer and tends to minimize the liability of the designer of record. One of the most important concepts to take from this study is:

> *Most roofing failures*
> *Could be eliminated*
> *By good peer reviews.*

Use systems with a successful track record. Many designers have gotten into problems by not checking on the performance history of a proposed system. Not too long ago, we checked with the owner or maintenance person at five locations about the performance of a product with which we were unfamiliar. The locations were all listed in the supplier's literature. In one case, they had never heard of the product. In three cases the product had failed and had been removed. In one case a law suit threatened. We are glad we did not use the product.

My own rather arbitrary limit for prior exposure is five years, with no problems in the same geographic area as the work under consideration. If a system gives unblemished service for five years, it will probably perform effectively for ten years. Obviously I would like to see more than five years exposure, but five years exposure is my minimum.

Minimize the roofing system failure by sticking to systems that work.

New and untested
Roofing systems generate
Consultants' income.

Monitoring the roofing work improves the system installed. The application can be improved by setting a neophyte on the roof wearing a monitor labeled hard hat. At least the work will improve until the roofing crew discovers the neophyte's lack of knowledge. A truly good monitor smoothes the way for the work, keeps track of all the problems and their remediation, encourages and confirms the job progress, and generally promotes the successful completion of the work.

Inexperienced monitors can interfere with job progress and are detested by the contractor. They can be easily identified by their rigid and unreasonable demands that hamper the work. For example, one monitor required the contractor to leave the edge of the roof without flashing until his supervisor could inspect the work several days later. That is nonsense. The flashing should be completed at the same time as the work in the field of the roof, to protect the system from moisture intrusion.

On the other hand, we have had clients who complained bitterly when we wanted to give a monitor a needed break so he could perform some of his personal business; they relied on his performance to expedite the work and yet maintain the high quality of the work we demand. I had one general contractor tell me, at the conclusion of a job: "You're a pain in the ass like all monitors, but you're a helpful pain in the ass."

Improve your roofing
Application and result
Through monitoring.

Buy by competence not price. It is almost axiomatic that problems will arise when a purchasing agent becomes involved. Too often, they want to buy by price rather than value or competence.

Several years ago a client had his purchasing agent hire a roofer to cut samples for our field investigation. We halted the proceedings when the man who showed up was only equipped with a penknife and drunk. We hired a roofing contractor known to us.

Our client noted quite a difference as three men set up a ladder, tied it off, and presented themselves on the roof fully equipped to cut and patch the roof. I doubt that our contractor even cost more – after all the samples were taken and the holes patched.

Every member of your team must be competent to perform the task. If even one team member is incompetent, the result will probably not be what you envision.

> *Buy by competence!*
> *Your money will go bye bye*
> *If you buy by price.*

Stupidity is one area not yet covered. Ignorance can be helped through education, but stupidity is unvaryingly fatal to the work, and is the cause of a host of roofing failures. It is hard to quantify and harder to correct, because the stupid person's belief gains strength the further it departs from reality. The belief in the importance of a warranty is a symptom of stupidity. The belief that a miracle compound, or angel dust, can invigorate and prolong the life of any roofing system is wishful thinking and a sign of gross stupidity.

Of course, stupidity is not the sole province of those involved in roofing. Politicians and bureaucrats are often candidates. One of Werner Gumpertz's favorite sayings is:

> Wenn Dummheit täte weh, oh welch erbärmlich schrei'n,
> würd' in der ganzen Welt in allen Häusern sein.

Translated into a haiku it would be:

> *If stupidity*
> *Were painful you would hear the*
> *Screams from everywhere!*

> *The final chapter!*
> *Congratulations, for now*
> *The end of our song.*

Appendix A – Roofing trade names

Trade name	Supplier	Description
#114 R	Merchant & Evans Inc.	Fastener
#1340 (A710)	American Tar Company	Asphalt emulsion
#15	Certain Teed Corporation	Organic ply felt
#15	Warrior Roofing Manufacturing Inc.	Asphalt organic felt
#15 ASTM	Warrior Roofing Manufacturing Inc.	Asphalt organic felt
#15 UL	Warrior Roofing Manufacturing Inc.	Asphalt organic felt
#15 W	Warrior Roofing Manufacturing Inc.	Asphalt organic felt
#30	Warrior Roofing Manufacturing Inc.	Asphalt organic felt
#30 ASTM & Goldline & UL	Warrior Roofing Manufacturing Inc.	Asphalt organic felt
#30 Shake liner	Warrior Roofing Manufacturing Inc.	Asphalt organic felt
#30 Split Felt	Warrior Roofing Manufacturing Inc.	Asphalt organic felt
#30 W	Warrior Roofing Manufacturing Inc.	Asphalt organic felt
#3036	Karnak Corporation	Poly ply mat
#305	Merchant & Evans Inc.	Fastener
#306	Merchant & Evans Inc.	Fastener
#31	Karnak Corporation	Glass ply sheet
#34	Karnak Corporation	Asphalt cotton fabric
#350	Malarkey Roofing Company	Cap sheet
#500	Malarkey Roofing Company	Glass ply sheet
#501	Malarkey Roofing Company	SBS Glass base
#502	Malarkey Roofing Company	Glass ply sheet
#506	Malarkey Roofing Company	Glass ply sheet
#515	Malarkey Roofing Company	Glass base sheet
#5548	Karnak Corporation	Resat Mat
#604	Henry Company	Glass base sheet
#605	Henry Company	80# Mineral surfaced underlayment
#606	Henry Company	SBS Glass base sheet
#607	Henry Company	Glass base sheet
#608	Henry Company	SBS Glass base sheet
20/20	Santoft Roof Tiles	Flat pan and cover clay tiles
2040M	Koppers Industries Inc.	APP – polyester granule surfaced

Appendix A (Continued)

Trade name	Supplier	Description
2040S	Koppers Industries Inc.	APP – polyester
2041M	Koppers Industries Inc.	SBS – polyester granule surfaced
2041MFR	Koppers Industries Inc.	SBS – polyester granule surfaced
2041S	Koppers Industries Inc.	SBS – polyester
2045M	Koppers Industries Inc.	SBS – polyester granule surfaced
2045MFR	Koppers Industries Inc.	SBS – polyester granule surfaced
2-Square #30 ASTM Felt F25	Fields Company	Asphalt organic felt
2-Square #30 Felt F22	Fields Company	Asphalt organic felt
3-square #15 ASTM Felt F35	Fields Company	Asphalt organic felt
3-square #15 Felt F30	Fields Company	Asphalt organic felt
3-square #15 Felt F32	Fields Company	Asphalt organic felt
4-square #15 Felt F40	Fields Company	Asphalt organic felt
4-square #15 Felt F42	Fields Company	Asphalt organic felt
525b – 2.5	SWD Urethane Company	PUF
525b – 3.0	SWD Urethane Company	PUF
5615 Base	Quantum Coatings, Inc.	Urethane coating
5615 Top	Quantum Coatings, Inc.	Urethane coating
7.2	MBCI	Galvalume
Accutrac Fastener	ITW Buildex	Screw – insulation to steel
Accutrac Fastener	ITW Buildex	Screw – insulation to wood
Acrylex 300	United Coatings	Acrylic
Acrylic Base	Metacrylics	Coating
Acrylic Beige	Metacrylics	Coating
Acrylic Brick Red	Metacrylics	Coating
Acrylic Custom Colors	Metacrylics	Coating
Acrylic Desert Sand	Metacrylics	Coating
Acrylic Gel	Metacrylics	Mastic
Acrylic Gray	Metacrylics	Coating
Acrylic Primer	Metacrylics	Adhesive
Acrylic Storm Cloud	Metacrylics	Coating
Acrylic White	Metacrylics	Coating
Adhere-It EPDM Primer	United Coatings	Urethane
Adhesive #9	Southwest Petroleum Company	Asphalt cement
Advanta Shingle	Atas International Inc.	Metal simulated shingle

AF Flashing Cement #19	Karnak Corporation	Asphalt cement
Alaskan #230 SBS Modified	Malarkey Roofing Company	Glass fiber – 3 tab strip shingle
ALCO Shield Water Prot. 100	ALCO-NVC	PMA pressure sensitive
ALCO Shield Water Prot. 195	ALCO-NVC	PMA pressure sensitive
ALCO Shield Water Prot. 200	ALCO-NVC	PMA pressure sensitive
ALCO Shield Water Prot. 225	ALCO-NVC	PMA pressure sensitive
Altusa S	Altusa/Intec, Corporation	Barrel mission clay tile
Alumagard Fib. Ctg. #215AF	ALCO-NVC, Inc.	Asphalt roof coating
Alumagard Non-Fib. Ctg. #214	ALCO-NVC, Inc.	Asphalt roof coating
Alumamation 301	Republic Powdered Metals	Asphalt coating
Alumaseal Primer	United Coatings	Urethane
Alumin-R Elast. Al #298 AF	Karnak Corporation	Coating
Aluminum #169 AF NF	Karnak Corporation	Asphalt coating
Aluminum Fibrated 2.0#	ALCM	Asphalt roof coating
Aluminum Fibrated 3.0#	ALCM	Asphalt roof coating
Aluminum Non-Fibrated 2.0#	ALCM	Asphalt roof coating
Aluminum Roof Coating	Southwest Petroleum Company	Asphalt coating
Aluminum Roof Shield	Southwest Petroleum Company	Asphalt coating
Americana	Ludowici Roof Tile Inc.	Flat clay tile
Amphibokote Wet/ Dry #155 AF	Karnak Corporation	Asphalt cement
Anchorbond #12	Celotex Corporation	Screw fastener
Anchorbond #12	Celotex Corporation	Screw fastener
Anchorbond #14	Celotex Corporation	Screw fastener
Anchorbond #14	Celotex Corporation	Screw fastener
Anchorbond #14	Celotex Corporation	Screw fastener
Anchorbond #14	Celotex Corporation	Screw fastener
Anchorbond #14 Stainless Steel	Celotex Corporation	Screw fastener
Anchorbond #14 Stainless Steel	Celotex Corporation	Screw fastener
Anchorbond #15 Heavy Duty	Celotex Corporation	Screw fastener
Anchorbond #15 Heavy Duty	Celotex Corporation	Screw fastener
Anchorbond #15 Heavy Duty	Celotex Corporation	Screw fastener

Appendix A (Continued)

Trade name	Supplier	Description
Anchorbond Augur Fastener	Celotex Corporation	Plastic screw
Antique	Ludowici Roof Tile Inc.	Flat clay tile
APOC 100 Plastic Cement	Gardner/APOC	Asphalt cement
APOC 101 Plastic Cement	Gardner/APOC	Asphalt cement
APOC 102 Plastic Cement	Gardner/APOC	Asphalt cement
APOC 103 Asphalt Primer	Gardner/APOC	Asphalt primer
APOC 104 Plastic Cement	Gardner/APOC	Asphalt cement
APOC 107 Fibre Cold-Ply	Gardner/APOC	Asphalt cement
APOC 109 Wet/Dry Cement	Gardner/APOC	Asphalt cement
APOC 122 Flashing Cement	Gardner/APOC	Asphalt cement
APOC 124 Wet/Dry Cement	Gardner/APOC	Asphalt cement
APOC 128 Flashing Cement	Gardner/APOC	Asphalt cement
APOC 133 MBA Flashing Cement	Gardner/APOC	PMA coating
APOC 136 MBA Adhesive	Gardner/APOC	PMA coating
APOC 211 Al Paint	Gardner/APOC	Asphalt coating
APOC 212 Al Coating	Gardner/APOC	Asphalt coating
APOC 252 Elastomeric White	Gardner/APOC	Coating
APOC 300 Asphalt Emulsion	Gardner/APOC	Asphalt emulsion
APOC 302 Fibered Emulsion	Gardner/APOC	Asphalt emulsion
APOC 337 Elastomeric Emulsion	Gardner/APOC	Asphalt emulsion
Aqua-lum Al Emulsion #297 AF	Karnak Corporation	Asphalt coating
AR Elastomeric #229 AF	Karnak Corporation	Coating
Architect 80/Estate	Certain Teed Corporation	Glass fiber – lam. metric shingle
Ardox H.T. Galvanize Concrete	National Nail Corporation	Friction fastener
Aristocrat 25	IKO Manufacturing Inc.	Organic – 3 tab strip shingle

Armour Lock 20	IKO Manufacturing Inc.	Organic shingle
Armour Plus 20	IKO Manufacturing Inc.	Organic – 3 tab strip shingle
Artic-Seal #170	Malarkey Roofing Company	PMA pressure sensitive
Asphalt 20	Georgia Pacific	Organic – 3 tab strip shingle
Asphalt Primer	ALCM	Asphalt primer
Asphalt Primer #108	Karnak Corporation	Asphalt primer
Asphalt Roof Primer #207	ALCO-NVC, Inc.	Asphalt primer
Asphalt Saturated Felt #15	Atlas Roofing Corporation	Asphalt organic felt
Asphalt Saturated Felt #30	Atlas Roofing Corporation	Asphalt organic felt
ASTM AR Heritage 25	Tamko Roofing Products	Glass fiber – rand. tab – lam. shingle
ASTM Heritage 25	Tamko Roofing Products	Glass fiber – rand. tab– lam. shingle
Atcobond #1822 (A200)	American Tar Company	Asphalt or Coal-tar coating
Atcobrite #5000	American Tar Company	Acrylic latex
Atcocoat #1818 (A100)	American Tar Company	Asphalt or Coal-tar coating
Atcogard #1840 (A700)	American Tar Company	Asphalt emulsion
Atcogard 2 #1850 (A750)	American Tar Company	Asphalt emulsion
Atcolap #1825 (A110)	American Tar Company	Asphalt or Coal-tar coating
Atcomastic #1823	American Tar Company	Asphalt or Coal-tar coating
Atcoprime #1931 (A400)	American Tar Company	Asphalt primer
Atcoscreen #1857 (A690)	American Tar Company	Asphalt or Coal-tar coating
Atcoshield #1859 (A650)	American Tar Company	Asphalt or Coal-tar coating
Atcoshield 2 #1864 (A640)	American Tar Company	Asphalt or Coal-tar coating
Atcostop # 1326	American Tar Company	Asphalt primer
Atcowhite #4200	American Tar Company	Acrylic latex
Auburn Lite 500 Series	Auburn Company	Flat concrete tile
Auburn Tile Regular Weight	Auburn Company	Flat concrete tile
Barrel	Altusa/Intec, Corporation	Barrel mission clay tile
Barrel	Monier Lifetile	Flat concrete tile
Base Sheet #43	Atlas Roofing Corporation	Asphalt-coated organic felt
Base Sheet #75	GAF Materials Company	Glass base sheet

Appendix A (Continued)

Trade name	Supplier	Description
Base Sheet Fastener Assembly	Tremko Inc.	Steel
Base-Loc	Simplex	Carbon–nylon–glass
Base-Loc	SFS Stadler Inc.	Nylon
Beauvoise	Huguenot Fenal	Flat clay tile
Benchmark	Conklin Co., Inc.	Acrylic latex
Berkley Pan & Cordova Cover	Gladding McBean	Flat pan and cover clay tile
Berkley Pan & Cover	Gladding McBean	Flat pan and cover clay tile
Berm 500	United Coatings	Asphalt emulsion
Berm 600/United 600	United Coatings	Acrylic
Bermuda Flat	Monier Lifetile	Flat concrete tile
Black Hawk	Green River Log Sales Ltd	Cedar shakes
Blended Mission	Monier Lifetile	Barrel mission concrete tile
Blended Shingle	Monier Lifetile	Flat concrete tile
Brittany	Ludowici Roof Tile Inc.	Flat clay tile
Buildcoat	ANDEK Corporation	Asphalt coating
Bermuda Roof Panel	Berridge Manufacturing Company	Metal "Bermuda" plank
Butyl Lastic	ALCM	Asphalt roof coating
C100 Roof Coat	Fields Co., LLC	Asphalt coating
C200 Roofbond	Fields Co., LLC	Asphalt cement
C240 Tilebond	Fields Co., LLC	Asphalt cement
C250 Roofflash	Fields Co., LLC	Asphalt cement
C3	Cooley Engineered Membrane	PVC Elvaloy KEE – polyester
C3 TPO	Cooley Engineered Membrane	Polypropylene
C300 Roof Mastic	Fields Co., LLC	Asphalt cement
Calais	Ludowici Roof Tile Inc.	Flat clay tile
Cambridge 25	IKO Manufacturing Inc.	Glass fiber – laminated shingle
Cambridge 30	IKO Manufacturing Inc.	Glass fiber – laminated shingle
Cambridge 40	IKO Manufacturing Inc.	Glass fiber – laminated shingle
Canal 40	TFB Tile	Barrel mission clay tile
Canal 50	TFB Tile	Barrel mission clay tile
Cap-Seam (CS)	AEP-Span	Metal standing seam
Capstone	Elk	Glass fiber – rand. tab – lam. shingle
Carlsile ASAP	Carlsile Syntec Incorporated	Screw – insulation to steel
Carlsile ASAP	Carlsile Syntec Incorporated	Screw – insulation to wood
Carriage House Shangle	Certain Teed Corporation	Glass fiber – 4 tab laminated shingle

Castle Top	Atas International Inc.	Metal diamond shaped shingle
Cathedral	IKO Manufacturing Inc.	Glass fiber shingle
Cathedral XL	IKO Manufacturing Inc.	Organic – 3 tab strip shingle
CD-10	Johns Manville International	Shank expansion
Cedar Plus	Green River Log Sales Ltd	Cedar shakes
Cedarlite	Monier Lifetile	Flat concrete tile
Cee-Lock Panel	Berridge Manufacturing Company	Galvanized steel, galvalume, copper
Celadon Ceramic Slate	Ludowici Roof Tile Inc.	Flat clay tile
Celo 1.045	Celotex Corporation	EPDM black
Celo 1.045 Reinforced	Celotex Corporation	EPDM black – polyester
Celo 1.045 Reinforced	Celotex Corporation	EPDM white – polyester
Celo 1.060	Celotex Corporation	EPDM white
Celo 1.060	Celotex Corporation	EPDM black
Celo 1.060 FR	Celotex Corporation	EPDM black
Celo 1.060 Reinforced	Celotex Corporation	EPDM black – polyester
Celotex APP 4/M Cap Sheet	Celotex	APP – polyester granule surfaced
Celotex APP 4/S Cap Sheet	Celotex	APP – polyester
CF Tap Grip	True-Fast Corporation	Screw fastener
Chalet	Atlas Roofing Corporation	Glass fiber – 3 tab shingle – in./lb
Chateau	IKO Manufacturing Inc.	Organic shingle
Chateau	Monier Lifetile	Flat concrete tile
Chateau Slate	Classic Products Inc.	Metal simulated slate
Chemfoam Contour Taper Tile	Pacemaker Plastics	EPS – type I
Chemfoam Contour Taper Tile	Pacemaker Plastics	EPS – type II
Chemfoam Contour Taper Tile	Pacemaker Plastics	EPS – type IX
Chemfoam Contour Taper Tile	Pacemaker Plastics	EPS – type VIII
Classic	Ludowici Roof Tile Inc.	Flat clay tile
Classic (Metric) 20 Year Traditional	Owens Corning Fiberglas	Glass fiber – 3 tab strip shingle
Classic 100	Monier Lifetile	Flat concrete tile
Classic 20 Year Traditional	Owens Corning Fiberglas	Glass fiber – 3 tab strip shingle
Classic AR 20 Year Traditional	Owens Corning Fiberglas	Glass fiber – 3 tab strip shingle

Appendix A (Continued)

Trade name	Supplier	Description
Classic Capri	Monier Lifetile	Barrel mission concrete tile
Classic Horizon Shangle	Certain Teed Corporation	Glass fiber – 3 tab shingle – in./lb
Classic Mission	Monier Lifetile	Barrel mission concrete tile
Classic Shingle	Berridge Manufacturing Company	Metal simulated shingle
Classic Tapered Mission	M.C.A. Clay Tile	Barrel mission clay tile
Clay Max Twin S Lightweight	U.S. Tile Company	Flat clay tile
Claylite Lightweight S Tile	U.S. Tile Company	Barrel mission clay tile
Closed Shingle	Ludowici Roof Tile Inc.	Flat clay tile
CMR-24	Butler Roof Division	Galvanized steel, galvalume
Cordova Tapered Custom	Gladding McBean	Barrel mission clay tile
Coffeyville English	Ludowici Roof Tile Inc.	Flat clay tile
Coil Roofing Nails	Simplex	Nail – smooth shanked
Cold Adhesive Cement #78 AF	Karnak Corporation	Asphalt cement
Cold Process Adhesive	ALCM	Asphalt cement
Colonial	Ludowici Roof Tile Inc.	Flat clay tile
Colonial Slate	Monier Lifetile	Flat concrete tile
Colorklad System 1	Vincent Metals	Standing seam metal panels
Colorklad System 2	Vincent Metal Goods	Various metals
Colorklad System 3	Vincent Metal Goods	Various metals
Colorklad System 4	Vincent Metal Goods	Various metals
Colorklad System 5	Vincent Metal Goods	Various metals
Colorklad System I	Vincent Metal Goods	Various metals
COMPABASE FA-2T	Bitec Inc.	APP – non-woven polyester
COMPABASE FS-2H	Bitec Inc.	SBS – non-woven glass
COMPABASE FS-2H-FR	Bitec Inc.	SBS – non-woven glass
COMPABASE PS-2H	Bitec Inc.	SPS – spunbond polyester
COMPAFLASH BFS-2H	Bitec Inc.	SPS – spunbond polyester
Consosera 8 in.	Ludowici Roof Tile Inc.	Flat clay tile
Continental	Dura-Loc Roofing Systems	Metal simulated tile
Cordova	Gladding McBean	Barrel mission clay tile
Cornada	Westile, Littleton, CO	Barrel mission concrete tile

Cotswold	Gladding McBean	Flat clay tile
Country Manor Shake	Perfection	Metal simulated shake
Country Pan Style	Santoft Roof Tiles	Barrel mission clay tile
Country Shingle	Monier Lifetile	Flat concrete tile
Country Slate	Monier Lifetile	Flat concrete tile
Country Split Shake	Monier Lifetile	Barrel mission concrete tile
Country Split Shingle	Monier Lifetile	Flat concrete tile
Craftsman Series High Batten	MBCI	Galvalume
Craftsman Series Large Batten	MBCI	Galvalume
Craftsman Series LB Profile	MBCI	Standing seam metal panels
Craftsman Series SB Profile	MBCI	Standing seam metal panels
Craftsman Series HB Profile	MBCI	Standing seam metal panels
Craftsman Series Small Batten	MBCI	Galvalume
Crowne 30	IKO Manufacturing Inc.	Organic – 3 tab strip shingle
Crude	Ludowici Roof Tile Inc.	Flat clay tile
CT 20	Certain Teed Corporation	Glass fiber – 3 tab shingle – in./lb
Curved Flat Seam	Berridge Manufacturing Company	Galvanized steel, galvalume
Curveline	Curveline Inc.	Various metals
Custom Lok 25	Certain Teed Corporation	Organic – lock shingle
Custom Sealdown 30	Certain Teed Corporation	Organic – 3 tab strip shingle
Deckfast #12	Construction Fasteners Inc.	Screw fastener
Deckfast #12	Construction Fasteners Inc.	Screw fastener
Deckfast #14	Construction Fasteners Inc.	Screw fastener
Deckfast #14	Construction Fasteners Inc.	Screw fastener
Deckfast #14	Construction Fasteners Inc.	Screw fastener
Deckfast #14 Stainless Steel	Construction Fasteners Inc.	Screw fastener
Deckfast #14 Stainless Steel	Construction Fasteners Inc.	Screw fastener
Deckfast #15 Heavy Duty	Construction Fasteners Inc.	Screw fastener
Deckfast #15 Hi Strength	Construction Fasteners Inc.	Screw fastener
Deckfast #15 Hi Strength	Construction Fasteners Inc.	Screw fastener
Decra Shake	Decra Roof Systems/Tasman Roofing	Metal simulated shake
Decra Slate	Decra Roof Systems/Tasman Roofing	Metal simulated slate

Appendix A (*Continued*)

Trade name	Supplier	Description
Decra Tile	Decra Roof Systems/Tasman Roofing	Metal simulated tile
Deklite	Construction Fasteners Inc.	Plastic screw
Dekspike	Construction Fasteners Inc.	Shank compression fastener
Dens-Deck 3/8 in.	Firestone Building Products	Glass mat faced gypsum
Dens-Deck Roof Board 1/2 in.	G-P Gypsum Corporation	Glass mat faced gypsum
Dens-Deck Roof Board 1/4 in.	G-P Gypsum Corporation	Glass mat faced gypsum
Dens-Deck Roof Board 5/8 in.	G-P Gypsum Corporation	Glass mat faced gypsum
Double Duty Aluminum LV	Tremco, Inc	Coating
Double Notched	Westile, Littleton, CO	Flat concrete tile
Double Roman Series #2000	Westile, Littleton, CO	Barrel mission concrete tile
DP	True-Fast Corporation	Screw fastener
DP (drill point)	True-Fast Corporation	Screw fastener
Drill-Tec #12	US Intec	Screw fastener
Drill-Tec 11#14	US Intec	Screw fastener
Drill-Tec 11#15 (High Strength)	US Intec	Screw fastener
Drill-Tec II#12	US Intec	Screw fastener
Drill-Tec II#14	US Intec	Screw fastener
Drill-Tec II#15	US Intec	Screw fastener
Drill-Tec II ES	US Intec	Galvanized steel
Drill-Tec Lite Deck	US Intec	Nylon
Drill-Tec Tap Grip	US Intec	Friction fastener
Drill-Tec TL	US Intec	Nylon
Dublin Slate	Westile, Littleton, CO	Flat concrete tile
Dura-Seal #202	Malarkey Roofing Company	Glass fiber – 3 tab strip shingle
Dura-Seal #204	Malarkey Roofing Company	Glass fiber – 3 tab strip shingle
Duro-Last Screws #14	Duro Last Inc.	Screw fastener
Duro-Last Auger Fastener	Duro Last Inc.	Plastic screw
Duro-Last Concrete Nail	Duro Last Inc.	Screw fastener
Duro-Last Concrete Screw	Duro Last Inc.	Screw fastener
Dutch Seam MRD	Atas International Inc.	Standing seam metal panels
Dynasty	IKO Manufacturing Inc.	Glass fiber – laminated shingle

Eave and Valley Shield	Globe Building Materials Inc.	PMA pressure sensitive
Eaveguard Shingle Underlay	Monsey Bakor Div. of Henry Co.	PMA pressure sensitive
Ecolastic	Tremco, Inc	Asphalt emulsion
Economy Round Metal Cap B/B	Simplex	Barbed fastener
EHD #15	True-Fast Corporation	Screw fastener
EHD (Extra Heavy Duty) #15	True-Fast Corporation	Screw fastener
Elasticote	ALCM	Coating
Elasto-Brite #501	Karnak Corporation	Coating
Elasto-Brite M #505 AF	Karnak Corporation	Coating
Elastron 858	United Coatings	Butyl
Elite Glass-Seal AR	Tamko Roofing Products	Glass fiber – 3 tab strip shingle
Elite Glass-Seal	Tamko Roofing Products	Glass fiber – 3 tab strip shingle
Els	Tremco, Inc.	Asphalt cement
Emergency Mastic	Garland Co.	Asphalt cement
Emergency Repair Mastic	Tremco, Inc.	Asphalt cement
Energizer FR	Garland Co.	PMA coating
Energizer K Plus FR	Garland Co.	PMA coating
ERS 100	Ecology Roof Systems	Asphalt emulsion
ERS 200	Ecology Roof Systems	PMA cement
ERS 300A	Ecology Roof Systems	Asphalt coating
ERS 300T	Ecology Roof Systems	Coal-tar coating
ERS 301	Ecology Roof Systems	Asphalt primer
ERS 302	Ecology Roof Systems	Asphalt coating
ERS 304	Ecology Roof Systems	Asphalt coating
ERS 305	Ecology Roof Systems	Asphalt coating
ERS 306	Ecology Roof Systems	Asphalt cement
ERS 307A	Ecology Roof Systems	Asphalt coating
ERS 307T	Ecology Roof Systems	Coal-tar coating
ERS 308	Ecology Roof Systems	Asphalt coating
ERS 309	Ecology Roof Systems	PMA cement
ERS 315	Ecology Roof Systems	Asphalt cement
ES-45 Base Ply Fastener	ES Products	Membrane to lightweight concrete
ES-60 Base Ply Fastener	ES Products	Membrane to lightweight concrete
ES-90 Base Ply Fastener	ES Products	Membrane to lightweight concrete
Espana Mission 100 Series	Monier Lifetile	Barrel mission concrete tile
Eternit Roofing	Eternit, Inc.	Fiber-cement shingle
Eternit Roofing Slates Continental	Eternit, Inc.	Fiber-cement shingle
Eternit Roofing Slates English	Eternit, Inc.	Fiber-cement shingle
Eternit Roofing Slates Thrutone	Eternit, Inc.	Fiber-cement shingle

Appendix A (Continued)

Trade name	Supplier	Description
Eternit Slates Alterna	Eternit, Inc.	Fiber-cement shingle
Europa	Santoft Roof Tiles	Barrel mission clay tile
Everguard EGHD	Gaf Materials Corporation	Screw fastener
Extra Load Fastener HD	SFS Stadler Inc.	Screw fastener
Extra Load Fastener HD	SFS Stadler Inc.	Screw fastener
Extra Load Fastener HD	SFS Stadler Inc.	Screw fastener
F100 Powrcoat	Fields Co., LLC	Asphalt coating
F110 Powrlap	Fields Co., LLC	Asphalt coating
F150 Powerseal	Fields Co., LLC	Asphalt coating
F200 Powrbond	Fields Co., LLC	Asphalt cement
F300 Powrmastic	Fields Co., LLC	Asphalt cement
F400 Powrprime	Fields Co., LLC	Asphalt primer
F460 Waterstop	Fields Co., LLC	Asphalt primer
F540 500 Aluminum Coating	Fields Co., LLC	Asphalt coating
F600 Flamebloc	Fields Co., LLC	Asphalt coating
F630, F640, F650 Al Ctg.	Fields Co., LLC	Asphalt coating
F670 Moblshield	Fields Co., LLC	Asphalt coating
F700 Powrgard	Fields Co., LLC	Asphalt emulsion
F750 Powrgard 2	Fields Co., LLC	Asphalt emulsion
F880 Sungard	Fields Co., LLC	Asphalt emulsion
Falcon Foam	Falcon Foam, Div. of Atlas Roofing	EPS – type I
Falcon Foam	Falcon Foam, Div. of Atlas Roofing	EPS – type II
Falcon Foam	Falcon Foam, Div. of Atlas Roofing	EPS – type IX
Falcon Foam	Falcon Foam, Div. of Atlas Roofing	EPS – type VIII
Fastener Grade	Topcoat Inc. (GAF)	Synthetic rubber, acrylic
FE 303 Series 2.7	Foam Enterprises Inc.	PUF – Type III
FE 303 Series 3.0	Foam Enterprises Inc.	PUF – Type III
FE 303-2.0	Foam Enterprises Inc.	PUF – Type I
FE 314-3.0 Series 3.0	Foam Enterprises Inc.	PUF – Type III
FE 700 Series Adhesive	Foam Enterprises Inc.	PUF – Type I
FE Coat 1000 Series	Foam Enterprises Inc.	Acrylic coating
FE Coat 2000 Series	Foam Enterprises Inc.	Urethane coating
FE Coat 3000 Series	Foam Enterprises Inc.	Silicone coating
FE Coat 4000 Series	Foam Enterprises Inc.	Butyl coating
Feather-Stone Slate/ Shake	Westile, Littleton, CO	Flat concrete tile
Fesco Board	Johns Manville International	Perlite – isofoam – organic/glass facer
Fesco Board	Johns Manville International	Perlite, type I
Fiber Tap	Firestone Building Products	Fiberboard, type II, grade 2

Fiberbase	Tremco Incorporated	Fiberboard, type II, grade 2
Fiberbase HD	Tremco Incorporated	Fiberboard, type II, grade 2
Fibered Emulsion #220 AF	Karnak Corporation	Asphalt emulsion
Fiberglass	Johns Manville International	Asphalt kraft – glass fiber
Fibermat	Tremco, Inc.	Asphalt cement
Fibertite	Seaman Corporation	EIP/KEE – polyester
Fibertite FB	Seaman Corporation	EIP/KEE – polyester – fleece backed
Fibrated Aluminum #97	Karnak Corporation	Asphalt coating
Fibrated Aluminum #97 AF	Karnak Corporation	Asphalt coating
Fibrated Aluminum #98 AF	Karnak Corporation	Asphalt coating
Fibrated Liquid Roof Coating	ALCM	Asphalt roof coating
Field Lock Seam	Atlas International	Galvanized steel, aluminized steel
Filter Vent	Air Vent Inc.	Aluminum vent
Firefree Plus PMFC Rustic Shake	Re-Con Building Products Inc.	Cellulose-cement shingle
Fire-Halt	Certain Teed Corporation	Glass fiber – laminated metric shingle
Fire-Halt 2000	Certain Teed Corporation	Glass fiber – laminated metric shingle
Firescreen	Certain Teed Corporation	Glass fiber – 3 tab strip shingle
Firescreen Plus 2000	Certain Teed Corporation	Glass fiber – 3 tab strip shingle
Firestone All Purpose	Firestone Building Products	Screw fastener
Firestone All Purpose	Firestone Building Products	Screw fastener
Firestone Concrete Drive	Firestone Building Products	Shank compression fastener
Firestone Heavy Duty	Firestone Building Products	Screw fastener
Firestone Heavy Duty	Firestone Building Products	Screw fastener
Firestone Heavy Duty	Firestone Building Products	Screw fastener
Firestone Polymer Fastener	Firestone Building Products	Glass reinforced nylon fastener
Firefree Plus PMFC Quarry Slate	Re-Con Building Products Inc.	Cellulose-cement shingle
Flame Bloc #1897	American Tar Company	Asphalt or Coal-tar coating
Flashband	ANDEK Corporation	PMA pressure sensitive
Flashbond Primer	ANDEK Corporation	Asphalt primer

Appendix A (Continued)

Trade name	Supplier	Description
Flashing	Garland Co.	Asphalt cement
Flashing Cement #216 AF	ALCO-NVC, Inc.	PMA cement
Flashing Grade	Topcoat Inc. (GAF)	Synthetic rubber, acrylic
Flashing Grade Spray Formula	Topcoat Inc. (GAF)	Synthetic rubber, acrylic
Flashtite	ALCM	Asphalt cement
Flat Slab Shingle	Ludowici Roof Tile Inc.	Flat clay tile
Flemish	Santoft Roof Tiles	Barrel mission clay tile
Flex FB 100	Flex Membrane International	KEE Evaloy – polyester – fleece backed
Flex FB Elvaloy	Flex Membrane International	KEE Evaloy – polyester – fleece backed
Flex MF/F 50	Flex Membrane International	PVC – polyester
Flex MF/R 60	Flex Membrane International	PVC – polyester
Flex MF/R Elvaloy	Flex Membrane International	KEE Evaloy – polyester
Flex-Cap Plastic Cap AG/B	Simplex	Annular groved fastener
Flex-Cap Plastic Cap AG/EGYD	Simplex	Annular groved fastener
Flex-Cap Plastic Cap AG/RL	Simplex	Annular groved fastener
Flexglas	Certain Teed Corporation	Glass base sheet
Flexglas Premium Cap Sheet 960	Certain Teed Corporation	Cap sheet
Flexglas Base Sheet	Certain Teed	APP – polyester granule surfaced
Flexglas Premium Cap 960	Certain Teed	APP – polyester sand
Flexseal	Topcoat Inc. (GAF)	Synthetic rubber, acrylic
Flex-Shield "EM" Roof Coating	Southwest Petroleum Company	Coating
Flex-Shield EM Patching Compound	Southwest Petroleum Company	Mastic
Flex-Shield Primer	Southwest Petroleum Company	Asphalt emulsion
Flex-Shield Roof Coating	Southwest Petroleum Company	Asphalt emulsion
Flintglas Cap Sheet	Certain Teed Corporation	Cap sheet
Flintglas Ply Sheet Type IV	Certain Teed Corporation	Glass ply sheet
Flintglas Ply Sheet Type VI	Certain Teed Corporation	Glass ply sheet
Flintlastic FR Base Sheet	Certain Teed	APP – polyester granule surfaced
Flintlastic FR Cap	Certain Teed	APP – polyester granule surfaced
Flintlastic FR-PG	Certain Teed	APP – polyester granule surfaced
Flintlastic GMS	Certain Teed	APP – polyester granule surfaced

Flintlastic GMS Premium	Certain Teed	APP – polyester granule surfaced
Flintlastic GTA	Certain Teed	APP – polyester granule surfaced
Flintlastic GTA-FR	Certain Teed	APP – polyester granule surfaced
Flintlastic GTS	Certain Teed	APP – polyester granule surfaced
Flintlastic STA	Certain Teed	APP – polyester aluminum
Flintlastic STA Plus 5	Certain Teed	APP – polyester aluminum
Flintlastic FR-P	Certain Teed	APP – polyester granule surfaced
Flush Panel	Peterson Aluminum Company	Standing seam metal panels
Fluted Concrete Nail	Gaf Materials Corporation	Friction fastener
Fluted Concrete Nail	Olympic Manufacturing Group	Friction fastener
FM 75 Base Fastener	ES Products	Friction – BUR to lightweight concrete
FM-45 Base Ply Fastener	ES Products	Membrane to lightweight concrete
FM-60 Base Ply Fastener	ES Products	Membrane to lightweight concrete
FM90 Base Ply	SFS Stadler Inc.	Steel
FM-90 Base Ply Fastener	ES Products	Membrane to lightweight concrete
Foamglas Insulation	Pittsburgh Corning Corp.	Cellular glass
Foamular Thermapink 18	Carlisle Syntec Incorporated	XPS, type IV
Foamular Thermapink 25	Carlisle Syntec Incorporated	XPS, type VII
Foamular Thermapink 40	Carlisle Syntec Incorporated	XPS, type VII
Formular 1000	Owens Corning	XPS, type V
Formular 105	Owens Corning	XPS, type X
Formular 250	Owens Corning	XPS, type IV
Formular 400	Carlisle Syntec Incorporated	XPS
Formular 400	Owens Corning	XPS, type VI
Formular 404	Owens Corning	XPS, type VI
Formular 404 RB	Owens Corning	XPS, type VI
Formular 600	Carlisle Syntec Incorporated	XPS
Formular 600	Owens Corning	XPS, type VII
Formular 604	Owens Corning	XPS, type VII
Formular 604 RB	Owens Corning	XPS, type VII
Formular Durapink	Carlisle Syntec Incorporated	XPS, type V
FS-25	Bitec Inc.	SBS – non-woven glass
FS-2H-Plus	Bitec Inc.	SBS – non-woven glass
FS-40	Bitec Inc.	SBS – non-woven glass
Fungusbuster 25	Certain Teed Corporation	Glass fiber – 3 tab shingle – in./lb
Futura-Flex 510	Quantum Coatings, Inc.	Urethane coating
Futura-Flex 550	Quantum Coatings, Inc.	Urethane coating

Appendix A (*Continued*)

Trade name	Supplier	Description
Futura-Thane 17060	Quantum Coatings, Inc.	Urethane coating
Futura-Thane 5007	Quantum Coatings, Inc.	Urethane coating
Futura-Thane 516	Quantum Coatings, Inc.	Urethane coating
Futura-Thane 550	Quantum Coatings, Inc.	Urethane coating
Futura-Thane 5600	Quantum Coatings, Inc.	Urethane coating
Futura-Thane 5600	Quantum Coatings, Inc.	Urethane coating
Futura-Thane 5625	Quantum Coatings, Inc.	Urethane coating
Futura-Thane 5650	Quantum Coatings, Inc.	Urethane coating
GAF Al Emulsion	GAF Materials Company	Asphalt emulsion
GAF Al Roof Paint	GAF Materials Company	Asphalt coating
GAF Asphalt/ Concrete Primer	GAF Materials Company	Asphalt primer
GAF Jetblack Premium Flashing Cement	GAF Materials Company	Asphalt cement
GAF Premium Fibered Al Roof Coating	GAF Materials Company	Asphalt coating
GAF Weathercoat Emulsion	GAF Materials Company	Asphalt emulsion
Gaftite #12-11/ Everguard EGIN	GAF Materials Corporation	Screw fastener
Gaftite #12-11/ Everguard EGIN	GAF Materials Corporation	Screw fastener
Gaftite #12-11/ Everguard EGIN-SS	GAF Materials Corporation	Screw fastener
Gaftite #14-10/ Everguard EGHD	GAF Materials Corporation	Screw fastener
Gaftite #14-10/ Everguard EGHD	GAF Materials Corporation	Screw fastener
Gaftite #14-10/ Everguard EGHD SS	GAF Materials Corporation	Screw fastener
Gaftite 1#14-10 (C-Steel)	GAF Materials Corporation	Screw fastener
Gaftite Base Sheet Fastener	GAF Materials Corporation	Split body fastener
Gaftite CD-10/ Everguard Spike	GAF Materials Corporation	Shank compression fastener
Gaftite Lit-Deck	GAF Materials Corporation	Friction
Garla-Brite	Garland Co.	Asphalt coating
Garla-Flex	Garland Co.	PMA coating
Garla-Prime	Garland Co.	Asphalt primer
Garla-Shield	Garland Co.	Asphalt emulsion
Garvitop	Garland Co.	Asphalt coating
Geogard	Republic Powdered Metals	Urethane

Gerard Shake	Gerard Roofing Technologies	Metal simulated shake
Gerard Tile	Gerard Roofing Technologies	Metal simulated tile
Glaslock 25 Year Interlocking	Owens Corning Fiberglas	Glass fiber – shingle
Glass-Mastic	ALCM	Asphalt cement
Glass Master Alpine	Atlas Roofing Corporation	Glass fiber – 3 tab shingle – in./lb
Glassmaster 25	Atlas Roofing Corporation	Glass fiber – 3 tab shingle – in./lb
Glassmaster T-Loc	Atlas Roofing Corporation	Glass fiber – T lock shingle
Glass-Seal	Tamko Roofing Products	Glass fiber – 3 tab strip shingle
Glazed	Tuilerie de Aleonard	Flat clay tile
Glassmaster 20	Atlas Roofing Corporation	Glass fiber – 3 tab shingle – in./lb
Gold Rush Series Espana Mission	Monier Lifetile	Barrel mission concrete tile
Goxhill Handmade	Santoft Roof Tiles	Flat clay tile
Grace Ice and Water Shield	W.R. Grace & Co. – Conn.	W.R. Grace & Co. – Conn.
Grand Manor Shangle	Certain Teed Corporation	Glass fiber – lam. rand. tab shingle
Grand Sequoia	GAF Materials Corporation	Glass fiber – lam. rand. tab shingle
Granutile	Atas International Inc.	Metal simulated tile
Green River	Green River Log Sales Ltd.	Cedar shakes and shingle
Grip Mastic	Garland Co.	Acrylic
Grundy Asphalt BU 68 Resaturant	Henry Company	Asphalt coating
Grundy Fibrated Roof Mastic II	Henry Company	Asphalt coating
Guardian EPDM Primer	Southwest Petroleum Company	Acrylic
Guardian General Purpose Primer	Southwest Petroleum Company	Acrylic
Guardian Seamless Roof Coating	Southwest Petroleum Company	Coating
H Kool Kote 1929	SWD Urethane Company	Acrylic coating
H14	Huguenot Fenal	French Barrel clay tile
Hallmark Shangle	Certain Teed Corporation	Organic – 3 tab shingle
Handcrafted	Daniel Platt Ltd	Flat clay tile
Hanseat 35	Heisterholz	Flat clay tile
Hardened Do-All Loc-Nail	ES Products	Split shank – single ply to wood
Hardened Do-All Loc-Nail	ES Products	Split shank – BUR to gypsum
Hardishake Roofing	James Hardie Building Products Inc.	Cellulose-cement shingle
Hardishingle Roofing	James Hardie Building Products Inc.	Cellulose-cement shingle

Appendix A (Continued)

Trade name	Supplier	Description
Hardislate Roofing	James Hardie Building Products Inc.	Cellulose-cement shingle
Hatteras	Certain Teed Corporation	Glass fiber – 4 tab laminated shingle
HD #14	True-Fast Corporation	Screw fastener
HD (Heavy Duty) #14	True-Fast Corporation	Screw fastener
HD 14-10 Fastener	Carlsile Syntec Incorporated	Screw fastener
HD Drill Point Stainless Steel	True-Fast Corporation	Screw fastener
HD Gravel Roof Preserver (Asphalt)	Southwest Petroleum Company	Asphalt coating
HD Gravel Roof Preserver (Coal-tar)	Southwest Petroleum Company	Asphalt coating
Hearthstead	Certain Teed Corporation	Organic – 4 tab strip shingle
Heavy Duty #14	True-Fast Corporation	Screw fastener
Heavy Duty #14-10	Tremco Inc.	Screw fastener
Heavy Duty Patching Compound	Southwest Petroleum Company	Asphalt cement
Heavy Duty Primer	Southwest Petroleum Company	Asphalt primer
Heavy Duty Roof Coating	Southwest Petroleum Company	Asphalt coating
Henry # 105 Low VOC Water Based Primer	Henry Company	Asphalt primer
Henry # 508 Wet Patch Roof Cement N/A	Henry Company	Asphalt cement
Henry #100 Elastomulsion	Henry Company	Asphalt emulsion
Henry #103 Low VOC Primer	Henry Company	Asphalt primer
Henry #104 Asphalt Primer	Henry Company	Asphalt primer
Henry #107 Asphalt Emulsion	Henry Company	Asphalt emulsion
Henry #109 Liquid Roof	Henry Company	Asphalt emulsion
Henry #111 Insulbond	Henry Company	Asphalt emulsion
Henry #112 Metalshield Solvent Primer	Henry Company	Coating
Henry #120 Premium Al Roof Coating	Henry Company	Asphalt coating

Henry #203 Cold Applied Cement	Henry Company	Asphalt coating
Henry #204 Plastic Roof Cement	Henry Company	Asphalt cement
Henry #208 Wet Patch Roof Cement	Henry Company	Asphalt cement
Henry #209 Elastomeric Cement	Henry Company	PMA cement
Henry #210 Asphalt Roof Coating	Henry Company	Asphalt coating
Henry #220 Premium Al Fibrated Roof Ctg	Henry Company	Asphalt coating
Henry #222 Elastomeric Flashing Cement	Henry Company	PMA cement
Henry #229 Aluminum Emulsion	Henry Company	Asphalt emulsion
Henry #280 White Elastomeric	Henry Company	Coating
Henry #287 Solarflex White Coating	Henry Company	Coating
Henry #294 Low VOC Metal Seam Sealer	Henry Company	Coating
Henry #295 Metal Seam Sealer	Henry Company	Coating
Henry #307 Fibrated Asphalt Emulsion	Henry Company	Asphalt emulsion
Henry #403 Cold Applied Cement	Henry Company	Asphalt cement
Henry #504 Plastic Roof Cement N/A	Henry Company	Asphalt cement
Henry #505 Flashmaster	Henry Company	PMA cement
Henry #517-518 Metalshield Roof System	Henry Company	Coating
Henry #520 Fibrated Al Roof Coating	Henry Company	Asphalt coating
Henry #521 3# Fibrated Aluminum	Henry Company	Asphalt coating
Henry #869 Elastomeric Al Roof Ctg	Henry Company	Asphalt coating

Appendix A (Continued)

Trade name	Supplier	Description
Henry #902 Permanent Bond-Adhesive	Henry Company	PMA adhesive
Henry #903 MB High Solids Adhesive	Henry Company	PMA adhesive
Henry #905 Flashmaster N/A	Henry Company	PMA cement
Henry #906 Flashmaster Plus	Henry Company	PMA cement
Henry Rubberkote Base Coat	Henry Company	Coating
Henry Rubberkote Primer	Henry Company	Coating
Henry Rubberkote Top Coat	Henry Company	Coating
Heritage 25	Tamko Roofing Products	Glass fiber – rand. tab – lam. shingle
Heritage 25 AR	Tamko Roofing Products	Glass fiber – rand. tab – lam. shingle
Heritage 30	Tamko Roofing Products	Glass fiber – rand. tab – lam. shingle
Heritage 30 AR	Tamko Roofing Products	Glass fiber – rand. tab – lam. shingle
Heritage 40	Tamko Roofing Products	Glass fiber – rand. tab – lam. shingle
Heritage 40 AR	Tamko Roofing Products	Glass fiber – rand. tab – lam. shingle
Heritage M25	Tamko Roofing Products	Glass fiber – rand. tab – lam. shingle
Heritage M30	Tamko Roofing Products	Glass fiber – rand. tab – lam. shingle
Heritage M40	Tamko Roofing Products	Glass fiber – rand. tab – lam. shingle
Heritage Stormfighter AR	Tamko Roofing Products	Glass fiber – rand. tab – lam. shingle
Henry # 275-276 Metalshield	Henry Company	Coating
Hextra	ITW Buildex	Screw fastener
Hextra	ITW Buildex	Screw fastener
Hextra Plus	ITW Buildex	Screw fastener
Hextra Plus	ITW Buildex	Screw fastener
High Sierra	Certain Teed Corporation	Glass fiber – laminated metric shingle
High Build Reflective Roof Coating	Tremco, Inc.	Elastomeric coating
High Snap-On Standing Seam	Peterson Aluminum Company	Standing seam metal panels

Highlander 25 #271	Malarkey Roofing Company	Glass fiber – laminated shingle
Highlands AR Shangle	Certain Teed Corporation	Glass fiber – shingle – in./lb
Highload ASAP	Johns Manville International	Screw – single ply to steel
Highload Fastener	Johns Manville International	Screw – single ply to steel
Hip & Ridge	IKO Manufacturing Inc.	Glass fiber – 4 tab strip shingle
Homestead	Monier Lifetile	Flat concrete tile
HP Concrete Spike	Carlsile Syntec Incorporated	Shank fastener
HP Fastener	Carlsile Syntec Incorporated	Screw fastener
HP Fastener	Carlsile Syntec Incorporated	Screw fastener
HP Fastener	Carlsile Syntec Incorporated	Screw fastener
HP High Speed Lock Toggle Bolt	Carlsile Syntec Incorporated	Toggle bolt
HP Lightweight Fastener	Carlsile Syntec Incorporated	Plastic screw
HP Lite-Deck	Carlsile Syntec Incorporated	Screw fastener
HP NTB With and Without Wires	Carlsile Syntec Incorporated	Plastic screw
HP Purlin Fastener	Carlsile Syntec Incorporated	Screw fastener
HP Toggle Bolt	Carlsile Syntec Incorporated	Toggle bolt
HP Woodie	Carlsile Syntec Incorporated	Screw fastener
HP-X Fastener	Carlsile Syntec Incorporated	Screw fastener
HP-X Fastener	Carlsile Syntec Incorporated	Screw fastener
Hurricane #240 SBS Modified, Algae Resistant	Malarkey Roofing Company	Glass fiber – 3 tab strip shingle
IB Single Ply	I B Roof Systems	PVC – polyester
ICA Premium APP Mineral	ICA Inc.	APP – polyester gravel surfaced
ICA Premium APP Slate Flakes	ICA Inc.	APP – polyester slate
ICA Premium APP Smooth	ICA Inc.	APP – polyester
Ice & Water Guard	Pabco Roofing Products	PMA pressure sensitive
Image II	Metal Sales Manufacturing Company	Standing seam metal panels
Imperial Gentry 25	IKO Manufacturing Inc.	Glass fiber – 3 tab strip shingle
Imperial Glass	IKO Manufacturing Inc.	Glass fiber – 3 tab strip shingle
Impressions	Dura-Loc Roofing Systems	Metal simulated shingle
Independence Shangle	Certain Teed Corporation	Glass fiber – lam. rand. tab shingle
Imperial Seal	IKO Manufacturing Inc.	Organic – 3 tab strip shingle
Insulated Blanket	I B Roof Systems	PVC – polyester
Insulation Round Metal Cap AG/B	Simplex	Annular grooved fastener

Appendix A (Continued)

Trade name	Supplier	Description
Insulation Round Metal Cap AG/ EGYD	Simplex	Annular grooved fastener
Insulation Round Metal Cap AG/RL	Simplex	Annular grooved fastener
Insulation Round Metal Cap B/B	Simplex	Barbed fastener
Insulation Round Metal Cap B/EGYD	Simplex	Barbed fastener
Insulation Round Metal Cap B/EGYD	Simplex	Barbed fastener
Insulation Round Metal Cap B/RL	Simplex	Barbed fastener
Insulbase	Polyglass USA	APP – polyester
Insul-Board	Insul-Board, Inc.	EPS – type I
Insuldeck Loc-Nail	ES Products	Knee bend – BUR to wood fiber-cement
Insul-Fixx #12-11	SFS Stadler Inc.	Screw fastener
Insul-Fixx #14-10	SFS Stadler Inc.	Screw fastener
Insul-Fixx #14-10	SFS Stadler Inc.	Screw fastener
Insul-Lite	SFS Stadler Inc.	Nylon
Insul-Lock Insulation Adhesive	Garland Co.	Urethane
Insulroofing	Polyglass USA	APP – polyester
Insulroofing Granular	Polyglass USA	APP – polyester granule surfaced
Intec Modified Base Plus	U.S. Intec Inc.	SBS – glass fiber
Integral Batten	Petersen Aluminum Corp.	Galvanized steel, aluminum
Integral Standing Seam	Petersen Aluminum Corp.	Galvanized steel, aluminum
Integral Batten	Peterson Aluminum Company	Standing seam metal panels
Integral Standing Seam	Peterson Aluminum Company	Standing seam metal panels
Interlock 18	Merchant & Evans Inc.	Many metals
International Black .045	International Diamond Systems	EPDM – black
International Black .060	International Diamond Systems	EPDM – black
International Fire Retardant .060	International Diamond Systems	EPDM – black
International Reinforced .060	International Diamond Systems	EPDM – black – scrim
International Reinforced .045	International Diamond Systems	EPDM – black – scrim

Inul-Fixx #12-11	SFS Stadler Inc.	Screw fastener
Inul-Fixx #14-10	SFS Stadler Inc.	Screw fastener
Iron-Lok Toggle Bolt	GAF Materials Corporation	Carbon steel
Iron-Lok Toggle Bolt	Olympic Manufacturing Group	Carbon steel
Iso-1	Johns Manville International	Isocyanurate foam, type II
ISO 95+ Composite	Firestone Building Products	Fiberboard – isofoam – organic/glass mat
ISO 95+ Composite	Firestone Building Products	Perlite – isofoam – organic/glass mat
Isofast IF2-C-M	SFS Stadler Inc.	Screw fastener
Isofast IF2-IS	SFS Stadler Inc.	Screw fastener
Isofast IF2-M	SFS Stadler Inc.	Screw fastener
Isofast IGM	SFS Stadler Inc.	Screw fastener
Isofast IW-T-M	SFS Stadler Inc.	Screw fastener
ISOP 95+ Isocyanurate	Firestone Building Products	Isocyanurate foam, type II
Italian Pan & Cordova Cover	Gladding McBean	Flat pan and cover clay tiles
Italian Pan & Cover	Gladding McBean	Flat pan and cover clay tiles
Italian Pan & Berkley Cover	Gladding McBean	Flat pan and cover clay tiles
Jet 25	Certain Teed Corporation	Glass fiber – shingle – in./lb
K 2000	Heisterholz	Flat clay tile
K21	Heisterholz	Flat clay tile
King-Con	ITW Buildex	Friction fastener
La Gauloise	TFB Tile	Flat clay tile
Landmark 25	Certain Teed Corporation	Glass fiber – rand. tab shingle
Landmark 30	Certain Teed Corporation	Glass fiber – rand. tab shingle
Landmark 40	Certain Teed Corporation	Glass fiber – rand. tab shingle
Legacy 35 #272 SBS Modified	Malarkey Roofing Company	Glass fiber – laminated shingle
Legacy 35 #272 SBS Modified AR	Malarkey Roofing Company	Glass fiber – laminated shingle
Legend	Atlas Roofing Corporation	Glass fiber – 3 tab shingle – in./lb
Lifepine Shakes	Tamark Manufacturing LLC	Treated pine shakes
Lightguard	T-Clear Corp.	3/8 in. concrete – EPS
Lightweight Interlocking	Ludowici Roof Tile Inc.	Flat clay tile
Lincoln Glazed Shingle	Gladding McBean	Flat clay tile
Lincoln Interlocking	Gladding McBean	Flat clay tile
Lincoln S-Tile	Gladding McBean	Barrel mission clay tile

Appendix A (Continued)

Trade name	Supplier	Description
Liquid Fabric-Flashing Grade	Topcoat Inc. (GAF)	Synthetic rubber, acrylic
Lite Deck	Olympic Manufacturing Group	Hardened carbon steel
Lite Weight Concrete Fastener	ITW Buildex	Galvanized steel fastener
Lite Weight Concrete Fastener	ITW Buildex	Galvanized steel fastener
Loc-Seam 360	Architectural Metal System	Galvalume
Loc-Seam Panel	Architectural Metal System	Galvalume
Lokseam	MBCI	Galvalume
Lokseam	MBCI	Standing seam metal panels
Long Span Panels	Architectural Metal System	Galvalume
LWC CR Base Sheet Fastener	Johns Manville International	Steel fastener
M100 Rubrcoat	Fields Co., LLC	PMA coating
M150 Rubrseal	Fields Co., LLC	PMA coating
M200 Rubrbond	Fields Co., LLC	PMA coating
M300 Rubrmastic	Fields Co., LLC	PMA coating
M400 Rubrprime	Fields Co., LLC	Asphalt primer
M600 Firebloc	Fields Co., LLC	Asphalt coating
M620 Silvermastic	Fields Co., LLC	Alkyd resin
M630 Silvershield	Fields Co., LLC	Asphalt coating
M700 Rubrgard	Fields Co., LLC	Asphalt emulsion
M800 Rubrstar	Fields Co., LLC	PMA emulsion
M850 Polrshield	Fields Co., LLC	Acrylic latex
M860 Polrbrite	Fields Co., LLC	Acrylic latex
Marathon 20	IKO Manufacturing Inc.	Glass fiber – 3 tab strip shingle
Marathon 25	IKO Manufacturing Inc.	Glass fiber – 3 tab strip shingle
Marathon 30	IKO Manufacturing Inc.	Glass fiber – 3 tab strip shingle
Marquis Weather Max	GAF Materials Corporation	Glass fiber – 3 tab strip shingle
Masonry Round Metal Cap	Simplex	Friction fastener
Master Smooth	Tamko Roofing Products Inc.	Asphalt-coated organic felt
Maxama	McElroy Metal Inc.	Standing seam metal panels
MB Priming System	Topcoat Inc. (GAF)	Asphalt primer
MBA Adhesive, Brush	ALCM	PMA cement
MBA Adhesive, Trowel	ALCM	PMA cement
Medallion 1	McElroy Metal Inc.	Standing seam metal panels

Medallion 2	McElroy Metal Inc.	Standing seam metal panels
Medallion-Loc	McElroy Metal Inc.	Standing seam metal panels
Meridian	McElroy Metal Inc.	Standing seam metal panels
Met-Tile	Met-Tile, Inc.	Metal mission tile
MF 108 Flat	M.C.A. Clay Tile	Flat clay tile
Mineral Lap Coating	Garland Co.	Acrylic
Mini, Maxi Batton	Metal Sales Manufacturing Company	Standing seam metal panels
Mira Vista – Designer Metal	Owens Corning	Metal simulated shingle
Mira Vista Copper	Owens Corning	Metal simulated shingle
Mira Vista Shake	Owens Corning Fiberglas	Slate, clay, resin
Mira Vista Slate	Owens Corning Fiberglas	Slate, clay, resin
Miradri WP 100	TC Miradri	PMA pressure sensative
Miradri WP 200	TC Miradri	PMA pressure sensative
Miradri WP 300 HT	TC Miradri	W.R. Grace & Co. – Conn.
Mission	Westile, Littleton, CO	Flat concrete tile
Mission 14-1/4 in.	Ludowici Roof Tile Inc.	Barrel mission clay tile
Mission S	Monier Lifetile	Barrel mission concrete tile
Mission S Desert	Monier Lifetile	Barrel mission concrete tile
Modified Bitumen Adhesive #66 AF	Karnak Corporation	PMA cement
Modified Bitumen Adhesive #81 AF	Karnak Corporation	PMA cement
Moisture Guard Plus	Tamko Roofing Products Inc.	PMA pressure sensitive
Monarch Roof MRB	Atas International Inc.	Batten seam standing seam
Monsey MBA Gold Adhesive	Henry Company	PMA cement
Monument Historique	Tuilerie de Aleonard	Flat clay tile
MOP Granule 170	Bakor	SBS – polyester granule surfaced
MP-300	Topcoat Inc. (GAF)	Synthetic rubber, acrylic
N.T.B. Magnum With & Without Wires	Duro Last Inc.	Plastic screw fastener
Nailbase	Firestone Building Products	OSB – isocyanurate – organic/glass mat
Nailboard	Johns Manville International	OSB – isocyanurate – organic/glass mat
Nail-Fast	Tamko Roofing Products Inc.	Asphalt-coated organic felt
Nail-Tite Type A	ES Products	BUR to gypsum

Appendix A (Continued)

Trade name	Supplier	Description
Nail-Tite Type R	ES Products	BUR to gypsum
Natural Roofing Slate	North Country Slate	Natural roofing slate
N-C	Olympic Manufacturing Group/N.B.T.	Screw fastener
NCI System 591 – 2.5	North Carolina Foam Industries	PUF
NCI System 591 – 2.8	North Carolina Foam Industries	PUF
NCI System 692 – 2.5	North Carolina Foam Industries	PUF
NCI System 692 – 2.8	Foam Enterprises Inc.	PUF
Neoprene Coating	Gardner/APOC	Neoprene
Neoprene Flashing Cement	Gardner/APOC	Neoprene
Neoprene Pitch Pad Sealant	Gardner/APOC	Neoprene
Neoprene Rubber Roof Cement	ALCM	Neoprene cement
New Englander 25	IKO Manufacturing Inc.	Organic shingle
New Horizon Shangle	Certain Teed Corporation	Glass fiber – shingle – in./lb
No Fiber Roof Emulsion	ALCM	Asphalt emulsion
No. 15	Tamko Roofing Products Inc.	Asphalt organic felt
No. 15	Tamko Roofing Products Inc.	Asphalt organic felt
No. 15 UL	Tamko Roofing Products Inc.	Asphalt organic felt
No. 30	Tamko Roofing Products Inc.	Asphalt organic felt
No. 30 18 in. Felt	Tamko Roofing Products Inc.	Asphalt organic felt
No. 30 ASTM Shake Underlayment	Tamko Roofing Products Inc.	Asphalt organic felt
No. 30 UL	Tamko Roofing Products Inc.	Asphalt organic felt
Non Fibrated Emulsion #100 AF	Karnak Corporation	Asphalt emulsion
Norman	Ludowici Roof Tile Inc.	Flat clay tile
NP 180 GgT FR	Bakor	SBS – polyester granule surfaced
NP 180 gM	Bakor	SBS – polyester granule surfaced
NP 180 p/p	Bakor	SBS – non-woven polyester
NP 180 p/s	Bakor	SBS – non-woven polyester
NP 180 gT	Bakor	SBS – polyester granule surfaced
NP 18 gM FR	Bakor	SBS – polyester granule surfaced
NP 250 Gm	Bakor	SBS – polyester granule surfaced

NP 250 gM	Bakor	SBS – polyester granule surfaced
NP 250 Gt	Bakor	SBS – polyester granule surfaced
NP 250 gT	Bakor	SBS – polyester granule surfaced
NTB	Johns Manville International	Nylon to gypsum fastener
N.T.B. Magnum With & Without Wires	Olympic Manufacturing Group	Nylon fastener
Oakridge 25 AR Architectural	Owens Corning Fiberglas	Glass fiber – rand. tab – lam. shingle
Oakridge 25 Architectural	Owens Corning Fiberglas	Glass fiber – rand. tab – lam. shingle
Oakridge 30 AR Shadow Architectural	Owens Corning Fiberglas	Glass fiber – rand. tab – lam. shingle
Oakridge 30 Shadow Architectural	Owens Corning Fiberglas	Glass fiber – rand. tab – lam. shingle
Oakridge 40 AR Deep Shadow Architectural	Owens Corning Fiberglas	Glass fiber – rand. tab – lam. shingle
Oakridge 40 Deep Shadow Architectural	Owens Corning Fiberglas	Glass fiber – rand. tab – lam. shingle
Olympic Base Sheet Fastener	Olympic Manufacturing Group	Steel
Olympic CD 10	Olympic Manufacturing Group	Shank compression fastener
Olympic Fastener #12 (C-Steel)	Olympic Manufacturing Group/ N.B.T.	Screw fastener
Olympic Fastener #12-11 (S-Steel)	Olympic Manufacturing Group/ N.B.T.	Screw fastener
Olympic Fastener HD #14 (C-Steel)	Olympic Manufacturing Group/ N.B.T.	Screw fastener
Olympic Fastener HD #14 (C-Steel)	Olympic Manufacturing Group/ N.B.T.	Screw fastener
Olympic Fastener HD #14 (S-Steel)	Olympic Manufacturing Group/ N.B.T.	Screw fastener
Olympic Fastener HD #14 (S-Steel)	Olympic Manufacturing Group	Screw fastener
Olympic Fastener HD (C-Steel)	Olympic Manufacturing Group	Screw fastener
Olympic Fastener Std #12	Olympic Manufacturing Group	Screw fastener
Olympic Fastener Std #12 (S-Steel)	Olympic Manufacturing Group	Screw fastener
One Coat Aluminum	Tremco, Inc.	Asphalt coating
One Piece S	M.C.A. Clay Tile	Barrel mission clay tile
One Step	Topcoat Inc. (GAF)	Synthetic rubber, acrylic
Oriental	M.C.A. Clay Tile	Oriental clay tile

Appendix A (Continued)

Trade name	Supplier	Description
Original Round Metal Cap AG/B	Simplex	Annular grooved fastener
Original Round Metal Cap AG/EYD	Simplex	Annular grooved fastener
Original Round Metal Cap AG/RL	Simplex	Annular grooved fastener
Original Round Metal Cap B/B	Simplex	Barbed fastener
Original Round Metal Cap B/EGYD	Simplex	Barbed fastener
Original Round Metal Cap B/RL	Simplex	Barbed fastener
Original Timerline	GAF Materials Corporation	Glass fiber – laminated shingle
Pabco GG-20	Pabco Roofing Products	Glass fiber – 3 tab strip shingle
Pabco Premier 25 year	Pabco Roofing Products	Glass fiber – laminated shingle
Pabco Premier 30 year	Pabco Roofing Products	Glass fiber – laminated shingle
Pabco Premier 40 year	Pabco Roofing Products	Glass fiber – laminated shingle
Pabco SG-25	Pabco Roofing Products	Glass fiber – 3 tab strip shingle
Palm Beach Tapered Mission	Ludowici Roof Tile Inc.	Barrel mission clay tile
Patch & Seal #992	Somay Products	Elastomeric
Patrimony	Tuilerie de Aleonard	Flat clay tile
PC Snap On Seam/ Batten	Atas International Inc.	Metal panels
Perma Shake	Atas International Inc.	Metal simulated shake
Permaflex	Republic Powdered Metals	Asphalt emulsion
Perma-Gard FR 7419 Base Coat	Neogard	Urethane
Perma-Seal FG Metal Roof & Tile Underlayment	Monsey Bakor Div. of Henry Co.	PMA pressure sensitive
Perma-Seal PE Metal Roof & Tile Underlayment	Monsey Bakor Div. of Henry Co.	PMA pressure sensitive
Phalempin	Huguenot Fenal	Flat clay tile
Pinnacle 30	Atlas Roofing Corporation	Glass fiber – laminated shingle – metric
Pinnacle 30	Atlas Roofing Corporation	Glass fiber – laminated shingle – in./lb

Pinnacle 40	Atlas Roofing Corporation	Glass fiber – laminated shingle – metric
Plain #30	Globe Building Materials Inc.	Asphalt organic felt
Plain Felt – UL #15	Certain Teed Corporation	Asphalt organic felt
Plain Felt – UL #30	Certain Teed Corporation	Asphalt organic felt
Plain Felt #15	Globe Building Materials Inc.	Asphalt organic felt
Plastic Cement	ALCM	Asphalt cement
Plasti-Cap	National Nail Corporation	Screw fastener
Plasti-Top	National Nail Corporation	Screw fastener
Plus Adhesive #269 AF	ALCO-NVC, Inc.	PMA cement
Polaroof AC	ANDEK Corporation	Acrylic latex
Poly Roof SF	Tremco, Inc.	Asphalt cement
Polymer Gyptec	ITW Buildex	Glass filled nylon fastener
Polyroof	Tremco, Inc.	Elastomeric cement
Power Rawl Speed-Lock Toggle	Powers Fasteners, Inc.	Screw fastener
Powers Rawl #12 Deck Screw	Powers Fastening Inc.	Screw fastener
Powers Rawl #12 Deck Screw	Powers Fasteners, Inc.	Screw fastener
Powers Rawl #14 Deck Screw	Powers Fastening Inc.	Screw fastener
Powers Rawl #14 Deck Screw	Powers Fasteners, Inc.	Screw fastener
Powers Rawl #14 Deck Screw	Powers Fastening Inc.	Screw fastener
Powers Rawl #14 Speed-Lock Toggle	Powers Fastening Inc.	Screw fastener
Powers Rawl #15 Deck Screw	Powers Fastening Inc.	Screw fastener
Powers Rawl #15 Deck Screw	Powers Fasteners, Inc.	Screw fastener
Powers Rawl #15 Deck Screw	Powers Fastening Inc.	Screw fastener
Powers Rawl 1/4" Spike	Powers Fastening Inc.	Shank compression fastener
Powers Rawl 3/16" Spike	Powers Fastening Inc.	Shank compression fastener
Powers Rawl Powerlite	Powers Fasteners	Nylon fastener
Powers Rawl Woodie	Powers Fastening Inc.	Screw fastener
Powers Rawl-Speed-Lock Toggle	Powers Fasteners	Carbon & stainless steel
Pre-Assembled Fastener	Firestone Building Products	Screw fastener
Pre-Assembled Fastener	Firestone Building Products	Screw – insulation to steel

Appendix A (Continued)

Trade name	Supplier	Description
Premier 25 Algae Block	Pabco Roofing Products	Glass fiber – laminated shingle
Premier 30 Algae Block	Pabco Roofing Products	Glass fiber – laminated shingle
Premier Advantage	Pabco Roofing Products	Glass fiber – laminated shingle
Premium #15	Globe Building Materials Inc.	Asphalt organic felt
Premium 25	Georgia Pacific	Glass fiber – 3 tab strip shingle
Premium Duralite Shake	Monier Lifetile	Flat concrete tile
Premium Duralite Villa	Monier Lifetile	Barrel mission concrete tile
Premium Fibered Aluminum Ctg #1866	American Tar Company	Asphalt or Coal-tar coating
Prestique II MD	Elk	Glass fiber – rand. tab – lam. shingle
Prestique I High Definition	Elk	Glass fiber – rand. tab – lam. shingle
Prestique II Raised Profile	Elk	Glass fiber – rand. tab – lam. shingle
Prestique Plus High Definition	Elk	Glass fiber – rand. tab – lam. shingle
Premium Flat Felt	Atlas Roofing Corporation	Asphalt organic/glass fiber felt
Prime & Seal	Somay Products	Acrylic
Pro Grade #842 A/F Fibrated Al Roof Ctg	Henry Company	Asphalt coating
Pro Asphalt Emulsion (fiber)	Dewitt Products Co.	Asphalt emulsion
Pro Asphalt Emulsion (no fiber)	Dewitt Products Co.	Asphalt emulsion
Pro Brite 200 Aluminum Fibre Coating	Dewitt Products Co.	Asphalt coating
Pro Coat Fiber Roof Coating	Dewitt Products Co.	Asphalt coating
Pro Flash Flashing Cement	Dewitt Products Co.	Asphalt cement
Pro Flash Wet/Stick Flashing Cement	Dewitt Products Co.	Asphalt cement
Pro Flash Xtra Flashing Cement	Dewitt Products Co.	PMA cement
Pro Grade #111 Asphalt Primer	Henry Company	Asphalt primer
Pro Grade #113 Asphalt Primer	Henry Company	Asphalt primer

Pro Grade #225 A/F All Weather Cement	Henry Company	Asphalt cement
Pro Grade #229 MB Flashing Cement	Henry Company	PMA cement
Pro Grade #25 All weather Roof Cement	Henry Company	Asphalt cement
Pro Grade #26 Plastic Roof Cement	Henry Company	Asphalt cement
Pro Grade #27 Flashing Cement	Henry Company	Asphalt cement
Pro Grade #31 Cold Process Adhesive	Henry Company	Asphalt coating
Pro Grade #331 A/F Cold Process Adhesive	Henry Company	Asphalt coating
Pro Grade #333 A/F MB Adhesive	Henry Company	PMA cement
Pro Grade #550 Elastomeric White Roof Ctg	Henry Company	Coating
Pro Grade #832 Unfibrated Al Roof Coating	Henry Company	Asphalt coating
Pro Lap Cement	Dewitt Products Co.	Asphalt cement
Pro Primer Asphalt	Dewitt Products Co.	Asphalt primer
Pro Resaturant Asphalt	Dewitt Products Co.	Asphalt primer
Pro Rooflox 300 Aluminum Fibre Coating	Dewitt Products Co.	Asphalt coating
Pro SBS Adhesive	Dewitt Products Co.	PMA cement
Pro SBS Flashing Cement	Dewitt Products Co.	PMA cement
Pro Silver shield 200 Al Ctg no fibre	Dewitt Products Co.	Asphalt coating
Pro Silverfibre Shield 300 Al Ctg	Dewitt Products Co.	Asphalt coating
Pro-Loc I, II, III	Metal Sales Manufacturing Company	Standing seam metal panels
Prominence 30 Year Performance	Owens Corning Fiberglas	Glass fiber – 3 tab strip shingle
Prominence AR 30 Year Performance	Owens Corning Fiberglas	Glass fiber – 3 tab strip shingle
Pro-Panel II	Metal Sales Manufacturing Company	Corrugated metal panel
Provincial	Ludowici Roof Tile Inc.	Flat clay tile

Appendix A (Continued)

Trade name	Supplier	Description
Prymsc	Garland Co.	Acrylic
Quantum Plus PMFC Shake Panel	Re-Con Building Products Inc.	Cellulose-cement shingle
Quantum Plus PMFC Slate Panel	Re-Con Building Products Inc.	Cellulose-cement shingle
R/S Round Top	National Nail Corporation	Screw fastener
Rainstop #1826 (A470)	American Tar Company	Asphalt or Coal-tar coating
RAM	Cooley Engineered Membrane	PVC Elvaloy KEE – polyester
Ram Tough Poly Felt 155 VP	Barrett Company	Poly ply felt
Ram Tough Poly Felt 265 VP	Barrett Company	Poly ply felt
Ram Tough Ram 203	Barrett Company	Glass roll roofing
Ram Tough Ram 30	Barrett Company	Poly base sheet
Ram Tough Ram 4	Barrett Company	Glass ply felt
Ram Tough Ram 6	Barrett Company	Glass ply felt
Ram Tough Ram Glass Cap Sheet	Barrett Company	Glass Cap Sheet
Ram Tough Ram Glass Mat	Barrett Company	Woven Glass ply sheet
Ram Tough Ram 32	Barrett Company	Glass base sheet
Rapid Roof II Base Coating	Conklin Co., Inc.	Acrylic coating
Rapid Roof III	Conklin Co., Inc.	Acrylic latex
Rapid Roof III Base Coating	Conklin Co., Inc.	Acrylic coating
R-Control Spec Lam	Pacemaker Plastics	OSB-EPS
R-Control Spec Lam	AMF R-Control	OSB-EPS-OSB
R-Control Spec Lam	Big Sky Insulations Inc.	OSB-EPS-OSB
R-Control Perform 1	Insulated Building Systems, Inc.	EPS – type I
R-Control Perform 1	Insulated Building Systems, Inc.	EPS – type XI
R-Control Perform 2 & 3	Insulated Building Systems, Inc.	Various with EPS
R-Control Perform Taper Tile	Insulated Building Systems, Inc.	EPS – type I
R-Control Perform Taper Tile	Insulated Building Systems, Inc.	EPS – type XI
R-Control SIP	AMF R-Control	OSB-EPS-OSB
R-Control SIP	Big Sky Insulations Inc.	OSB-EPS-OSB
R-Control SIP	Pacemaker Plastics	OSB-EPS-OSB
RC-W Elasto-Kote #502	Karnak Corporation	PMA cement
Recover Board 1/2 in.	GAF Materials Corporation	Perlite, type I

Redi-Roof Batten	Petersen Aluminum Corp.	Galvanized steel, aluminum
Redi-Roof Batten	Peterson Aluminum Company	Standing seam metal panels
Redi-Roof Standing Seam	Petersen Aluminum Corp.	Galvanized steel, aluminum
Redi-Roof Standing Seam	Peterson Aluminum Company	Standing seam metal panels
Regular Density Roof Insulation	Huebert Fiberboard Company	Fiberboard, type I
Regular Fiberboard	Koppers Industries Inc.	Fiberboard, type I
Reliance 25	Tamko Roofing Products	Organic – 3 tab strip shingle
Reliance 30	Tamko Roofing Products	Organic – 3 tab strip shingle
Renaissance XL	IKO Manufacturing Inc.	Organic – 3 tab strip shingle
Residence	TFB Tile	Barrel mission clay tile
Retrofit Board 1/2 in.	Johns Manville International	Perlite, type I
Reveal Panel	Peterson Aluminum Company	Standing seam metal panels
R-Mer Lite	The Garland Co., Inc.	Galvalume
Roma	Monier Lifetile	Barrel mission concrete tile
Roman	Ludowici Roof Tile Inc.	Flat clay tile
Roman Canal	TFB Tile	Barrel mission clay tile
Romane	Huguenot Fenal	Barrel mission clay tile
Romane	TFB Tile	Barrel mission clay tile
Roof Cap Mastic #270 AF	ALCO-NVC, Inc.	PMA cement
Roof Coating #71 AF	Karnak Corporation	Asphalt coating
Roof Emulsion Fibrated	ALCM	Asphalt emulsion
Roof Grip	ITW Buildex	Screw fastener
Roof Grip	ITW Buildex	Screw fastener
Roof Grip #14	ITW Buildex	Screw fastener
Roof Grip #15	ITW Buildex	Screw fastener
Roof Grip Plus	ITW Buildex	Screw fastener
Roof Grip Plus	ITW Buildex	Screw fastener
Roof Mastic #842	Somay Products	Elastomeric
Roof Shield 60	United Coatings	Acrylic
Roofers Select	Certain Teed Corporation	Asphalt organic/glass fiber felt
Roofmate	United Coatings	Acrylic
Rooftec CA	Intercontinental Coatings Corp.	Coal-tar, KEE, Evaloy – polyester
Rooftec HA	Intercontinental Coatings Corp.	Coal-tar, KEE, Evaloy – polyester
Rooftec SA	Intercontinental Coatings Corp.	Coal-tar, KEE, Evaloy – polyester

Appendix A (Continued)

Trade name	Supplier	Description
Rooftec SABV	Intercontinental Coatings Corp.	Coal-tar, KEE, Evaloy – polyester
Rooftec SABVWS	Intercontinental Coatings Corp.	Coal-tar, KEE, Evaloy – polyester
Rooftec WS	Intercontinental Coatings Corp.	Coal-tar, KEE, Evaloy – polyester
Rough Shake	Classic Products Inc.	Metal simulated shake
Round-Top Masonry	National Nail Corporation	Screw fastener
Royal Shingle	Tamark Manufacturing LLC	Treated pine shakes
Royal Sovereign	GAF Materials Corporation	Glass fiber – 3 tab strip shingle
Royal Victorian	IKO Manufacturing Inc.	Organic – random tab shingle
R-Panel	Berridge Manufacturing Company	Galvanized steel, galvalume
RPI EPDM Black .045	Roofing Products International	EPDM – black
RPI EPDM Black .045	Roofing Products International	EPDM – black – polyester
RPI EPDM Black .060	Roofing Products International	EPDM – black
RPI EPDM Black .060	Roofing Products International	EPDM – black – polyester
RPI EPDM Black FR .045	Roofing Products International	EPDM – black
RPI EPDM Black FR .060	Roofing Products International	EPDM – black
RPI EPDM White .045	Roofing Products International	EPDM – white
RPI EPDM White .060	Roofing Products International	EPDM – white
Rubbergard .045	Firestone Building Products	EPDM – black
Rubbergard .045 FR	Firestone Building Products	EPDM – black
Rubbergard .045 LSFR	Firestone Building Products	EPDM – black
Rubbergard .060	Firestone Building Products	EPDM – black
Rubbergard .060 FR	Firestone Building Products	EPDM – black
Rubbergard .060 LSFR	Firestone Building Products	EPDM – black
Rubbergard .090	Firestone Building Products	EPDM – black
Rubbergard Max .045 Reinforced	Firestone Building Products	EPDM – black – polyester
Rubbergard Max .060 Reinforced	Firestone Building Products	EPDM – black – polyester
Rubberized Cement	ALCM	Asphalt cement
Rubberized Damp Surface Roof Coating	ALCM	Asphalt roof coating

Ruberglas 2 M66	Fields Company, LLC	Glass base sheet
Ruberglas M62	Fields Company, LLC	Glass base sheet
Ruberoid 20	GAF Materials Corporation	SBS – non-woven glass
Ruberoid 30	GAF Materials Corporation	SBS – glass fiber granule surfaced
Ruberoid 30 FR	GAF Materials Corporation	SBS – glass fiber granule surfaced
Ruberoid Modified Bitumen Adhesive	GAF Materials Company	PMA adhesive
Ruberoid Modified Bitumen Flashing Ad.	GAF Materials Company	PMA coating
Ruberoid Mop 170 FR	GAF Materials Corporation	SBS – polyester granule surfaced
Ruberoid Mop FR	GAF Materials Corporation	SBS – polyester granule surfaced
Ruberoid Mop Granule	GAF Materials Corporation	SBS – polyester granule surfaced
Ruberoid Mop Plus	GAF Materials Corporation	SBS – polyester granule surfaced
Ruberoid Mop Smooth	GAF Materials Corporation	SBS – polyester
Ruberoid SBS HW FR	GAF Materials Corporation	SBS – polyester granule surfaced
Ruberoid SBS HW Granule	GAF Materials Corporation	SBS – polyester granule surfaced
Ruberoid SBS HW Plus	GAF Materials Corporation	SBS – polyester granule surfaced
Ruberoid SBS HW Plus FR	GAF Materials Corporation	SBS – polyester granule surfaced
Ruberoid SBS HW Smooth	GAF Materials Corporation	SBS – polyester
Ruberoid Torch 1	GAF Materials Corporation	APP – polyester granule surfaced
Ruberoid Torch FR	GAF Materials Corporation	APP – polyester granule surfaced
Ruberoid Torch Granule	GAF Materials Corporation	APP – polyester granule surfaced
Ruberoid Torch Plus	GAF Materials Corporation	APP – polyester granule surfaced
Ruberoid Torch Smooth	GAF Materials Corporation	APP – polyester
Ruberpoly 42 – M64	Fields Company, LLC	Glass base sheet
Ruberpoly M60	Fields Company, LLC	Glass base sheet
Rust-Go	Garland Co.	Alkyd
Rustic Shake Shingle	Berridge Manufacturing Company	Metal simulated shake
Rustic Shingle	Classic Products Inc.	Metal simulated shake
RWP Roofing Panel	IPS Insulated Panel Systems	Galvanized steel, galvalume
S-Tile	U.S. Tile Company	Barrel mission clay tile
Santa Fe Mission	Monier Lifetile	Barrel mission concrete tile

Appendix A (Continued)

Trade name	Supplier	Description
Santa Fe Mission barrel	Santa Fe Tile Corporation	Barrel mission clay tile
Santa Fe Royal	Santa Fe Tile Corporation	Flat clay tile
Santa Fe S	Santa Fe Tile Corporation	Barrel mission clay tile
Santa Fe Shingle	Monier Lifetile	Flat concrete tile
Satellite Type 15 – #402	Malarkey Roofing Company	Asphalt organic felt
Satellite Type 30 – #404	Malarkey Roofing Company	Asphalt organic felt
Satellite Type 30 Shakeliner #405	Malarkey Roofing Company	Asphalt organic felt
SB-900 Flashing Grade	Topcoat Inc. (GAF)	Synthetic rubber, acrylic
SBM 18 in.	Ludowici Roof Tile Inc.	Barrel mission clay tile
SBS Trowel Grade #269t AF	ALCO-NVC, Inc.	PMA cement
SBS Underlayment #501 UDL	Malarkey Roofing Company	PMA glass fiber felt
Scanroof	Atas International Inc.	Metal simulated slate
Scott Cedar	Green River Log Sales Ltd	Cedar Shakes
Seal King 25	Certain Teed Corporation	Glass fiber – 3 tab shingle – in./lb
Sealdon 25	Certain Teed Corporation	Glass fiber – 3 tab shingle – in./lb
Senco Base Tape System	Senco Products, Inc.	Staple – BUR to wood
Sentinel	GAF Materials Corporation	Glass fiber – 3 tab strip shingle
Shadowline	Dura-Loc Roofing Systems	Metal simulated slate
Shake	Monier Lifetile	Flat concrete tile
Shake	Westile, Littleton, CO	Flat concrete tile
Shake	Dura-Loc Roofing Systems	Metal simulated shake
Shake Felt 18 in.	Globe Building Materials Inc.	Asphalt organic felt
Shake Underlayment 22 in.	Tamko Roofing Products Inc.	Asphalt organic felt
Shakeliner F10	Fields Company	Asphalt organic felt
Shakeliner F12	Fields Company	Asphalt organic felt
Shakeliner F15 Type 30 ASTM	Fields Company	Asphalt organic felt
Shakeliner F20 2-Square	Fields Company	Asphalt organic felt
Sheeting Bond Black	Tremco, Inc.	Elastomeric cement
Sheeting Bond White	Tremco, Inc.	Elastomeric cement
Shingle Felt	Warrior Roofing Manufacturing Inc.	Asphalt organic felt
Shingle Underlayment	Globe Building Materials Inc.	Asphalt organic felt
Shingle Vent II	Air Vent Inc.	HD polymer vent
Shingle-Mate	GAF Materials Corporation	Asphalt organic/glass fiber felt
Sierra Shake	Monier Lifetile	Flat concrete tile

Sierra Mission	Westile, Littleton, CO	Barrel mission concrete tile
Silver Shield	Garland Co.	Asphalt coating
Silverseal #1870 (A610)	American Tar Company	Asphalt or Coal-tar coating
Skyline 25	IKO Manufacturing Inc.	Organic shingle
Skylite	Topcoat Inc. (GAF)	Synthetic rubber, acrylic
Slate	Monier Lifetile	Flat concrete tile
Slate	Westile, Littleton, CO	Flat concrete tile
Slate/Tile Underlayment	Atlas Roofing Corporation	Asphalt-coated organic felt
Stateline	GAF Materials Corporation	Glass fiber – random tab shingle
Snap Rib SSR 3"	Fabral	Many metals
Snap-Clad	Peterson Aluminum Company	Standing seam metal panels
Snap-On Batten	Peterson Aluminum Company	Standing seam metal panels
Snap-On Standing Seam	Peterson Aluminum Company	Standing seam metal panels
Solargard	Republic Powdered Metals	Acrylic
Solargard Hy-Build	Republic Powdered Metals	Acrylic
Solargard Ultra	Republic Powdered Metals	Elastomeric
Spanish	Ludowici Roof Tile Inc.	Barrel mission clay tile
Spanish 18 in.	Ludowici Roof Tile Inc.	Barrel mission clay tile
Spanish S	Monier Lifetile	Barrel mission concrete tile
Spanish S-Tile	Gladding McBean	Barrel mission clay tile
Spanish Tile	Berridge Manufacturing Company	Metal simulated tile
Specification Felt #15	Atlas Roofing Corporation	Asphalt organic felt
Specification Felt #30	Atlas Roofing Corporation	Asphalt organic felt
Split Shake	Monier Lifetile	Flat concrete tile
Springhouse Shingles	Springhouse, Inc.	Metal simulated shingles/shakes
Square Head Metal Cap B/B	Simplex	Barbed fastener
SSC	ARS Industries	Snap on batten metal panels
SSR 1-1/2	Fabral	galvanized steel
SSR 2-1/2	Fabral	Many metals
Stadler Spike	SFS Stadler Inc.	Shank compression fastener
Stadler Spike	SFS Stadler Inc.	Shank compression fastener
Standard #12-11	Tremco Inc.	Screw fastener
Standing Seam Shingle	Atas International Inc.	Standing seam metal panels
S-Tile	Metal Sales Manufacturing Company	Metal simulated tile
S-Tile	Berridge Manufacturing Company	Metal simulated tile

Appendix A (Continued)

Trade name	Supplier	Description
Stormguard Waterproof Underlayment	GAF Materials Corporation	PMA pressure sensitive
Stormmaster DG Ice and Water Protection	Atlas Roofing Corporation	PMA pressure sensitive
Stormmaster LM SBS Modified	Atlas Roofing Corporation	Glass fiber – laminated shingle – in./lb
Stormmaster ST SBS Modified	Atlas Roofing Corporation	Glass fiber – 3 tab shingle – in./lb
Stormshield Black Diamond Base Sheet	Certain Teed Corporation	PMA pressure sensitive
Stormtite 25	IKO Manufacturing Inc.	Glass fiber shingle
Straight Barrel Mission	M.C.A. Clay Tile	Barrel mission clay tile
Summit	Georgia Pacific	Glass fiber – laminated shingle
Summit III	Georgia Pacific	Glass fiber – laminated shingle
Sunguard	Sunguard Marketing Corporation	Elastomeric
Sunguard R	Kokem Products Inc.	Coating
Sunshield #1858	American Tar Company	Asphalt or Coal-tar coating
Supershake	Monier Lifetile	Flat concrete tile
Supreme (Metric) 25 Year Traditional	Owens Corning Fiberglas	Glass fiber – 3 tab strip shingle
Supreme 25 Year Traditional	Owens Corning Fiberglas	Glass fiber – 3 tab strip shingle
Supreme 30 Year Performance	Owens Corning Fiberglas	Glass fiber – 3 tab strip shingle
Supreme AR 30 Year Performance	Owens Corning Fiberglas	Glass fiber – 3 tab strip shingle
Supreme AR 25 Year Traditional	Owens Corning Fiberglas	Glass fiber – 3 tab strip shingle
Sure Fast Fastener	Carlsile Syntec Incorporated	Screw fastener
Sure Fast Fastener	Carlsile Syntec Incorporated	Screw fastener
Surface Seal SB	Topcoat Inc. (GAF)	Synthetic rubber, acrylic
SV Crimp	Metal Sales Manufacturing Company	Metal panels
System ES I #12-11	SFS Stadler Inc.	Screw fastener
System ES I #12-11	SFS Stadler Inc.	Screw fastener
System ES I #14-10	SFS Stadler Inc.	Screw fastener
System ES I #14-10	SFS Stadler Inc.	Screw fastener
System ES L #14-10	SFS Stadler Inc.	Screw fastener
System Metal ES	SFS Stadler Inc.	Screw fastener

Tam-loc Glass	Tamko Roofing Products	Glass fiber – shingle
Tam-Pro Asphalt Primer	Tamko Roofing Products	Asphalt primer
Tam-Pro Cold Applied Cement	Tamko Roofing Products	Asphalt cement
Tam-Pro CPA-SBS Flashing Cement	Tamko Roofing Products	PMA cement
Tam-Pro CPA-SBS Adhesive	Tamko Roofing Products	PMA cement
Tam-Pro Fibered Emulsion Coating	Tamko Roofing Products	Asphalt emulsion
Tam-Pro Fibrated Al Roof Coating	Tamko Roofing Products	Asphalt coating
Tam-Pro Fibrated Roof Coating	Tamko Roofing Products	Asphalt coating
Tam-Pro Fire Rate FR Fiber Al Coating	Tamko Roofing Products	Asphalt coating
Tam-Pro Heavy Body Flashing Cement	Tamko Roofing Products	Asphalt cement
Tam-Pro Non-Fibered Al Coating	Tamko Roofing Products	Asphalt coating
Tam-Pro Plastic Roof Cement	Tamko Roofing Products	Asphalt cement
Tam-Pro Q15 Elastomeric Flashing Cement	Tamko Roofing Products	Asphalt cement
Tam-Pro Wet/Dry Plastic Cement	Tamko Roofing Products	Asphalt cement
Taos Mission	Monier Lifetile	Barrel mission concrete tile
Taos Shingle	Monier Lifetile	Flat concrete tile
Tapered Mission 16 in.	Ludowici Roof Tile Inc.	Barrel mission clay tile
Tapered Mission Tile	U.S. Tile Company	Barrel mission clay tile
Tar Cement #170 AF	Karnak Corporation	Coal-tar cement
Tech Tile	Atas International Inc.	Spanish metal panels
Thermo Shield Roof Coating	SPM Thermo-Shield, Inc.	Acrylic
Tile Underlayment	Tamko Roofing Products Inc.	Asphalt coated organic felt
Tilestone	Ludowici Roof Tile Inc.	Flat clay tile
Timberline 25	GAF Materials Corporation	Glass fiber – laminated shingle
Timberline Country Mansion	GAF Materials Corporation	Glass fiber – laminated shingle
Timberline Ultra	GAF Materials Corporation	Glass fiber – laminated shingle

Appendix A (Continued)

Trade name	Supplier	Description
TL (Tectum-Lightweight)	True-Fast Corporation	Nylon fastener
Toggle Bolt – carbon steel	GAF Materials Corporation	Carbon steel
Toggle Bolt – stainless steel	GAF Materials Corporation	Stainless steel
Toggle Bolt (carbon steel)	Olympic Manufacturing Group	Carbon steel
Toggle Bolt (stainless steel)	Olympic Manufacturing Group	Stainless steel
Toggle-less Magnum/Everguard	GAF Materials Corporation	Nylon fastener
Top S 010	Topcoat Inc. (GAF)	Synthetic rubber, acrylic
Topcoat CRT	Topcoat Inc. (GAF)	Synthetic rubber, acrylic
Topcoat MB Plus	Topcoat Inc. (GAF)	Synthetic rubber, acrylic
Topcoat Membrane	Topcoat Inc. (GAF)	Synthetic rubber, acrylic
Topcoat XR 2000	Topcoat Inc. (GAF)	Synthetic rubber, acrylic
TORCH Granule 170	Bakor	SBS polyester granule surfaced
Tough-Glass	Georgia Pacific	Glass fiber – 3 tab strip shingle
Tough-Glass Plus	Georgia Pacific	Glass fiber – 3 tab strip shingle
Tough-Glass T-Lock	Georgia Pacific	Glass fiber shingle
TP	True-Fast Corporation	Screw fastener
TP	True-Fast Corporation	Screw fastener
TPR – The Peel Rivit	SFS Stadler Inc.	Aluminum alloy
TPR – The Peel Rivit	SFS Stadler Inc.	Clamping fastener
Trem Prime QD Low Odor	Tremco, Inc.	Asphalt primer
Tremfix	Tremco, Inc.	Asphalt cement
Tremlar LRM-H	Tremco, Inc.	PMA cement
Tremlar LRM-V	Tremco, Inc.	PMA cement
Tremlastic	Tremco, Inc.	Asphalt emulsion
Tremlastic S	Tremco, Inc.	Asphalt emulsion
Tremlite Coating	Tremco, Inc.	Acrylic
Tremlite Epoxy Rust Coat Low Odor	Tremco, Inc.	Epoxy coating
Tremlite Mastic	Tremco, Inc.	Acrylic
Tremlite Metal Primer W.B.	Tremco, Inc.	Acrylic
Tremply W.B.	Tremco, Inc.	Asphalt primer
Tri Ply Asphalt Cement Primer	GAF Materials Company	Asphalt primer
Tri Ply Modified Bitumen Flashing Ad.	GAF Materials Company	PMA coating

Tri Ply Modified Bitumen Flashing Cement	GAF Materials Company	PMA coating
Tri Ply Premium Fibrated Aluminum	GAF Materials Company	Asphalt coating
Triton	Daniel Platt Ltd	Flat clay tile
Trulite Lightweight Concrete Tile	Monier Lifetile	Barrel mission concrete tile
Trulite Lightweight Concrete Tile	Monier Lifetile	Flat concrete tile
Tube-Lok B	Simplex	Carbon steel
Tube-Lok EGYD	Simplex	Carbon steel
Tube-Lok RL	Simplex	Carbon steel
Turret Tile	M.C.A. Clay Tile	Barrel mission clay tile
Twin Loc-Nail	ES Products	Staple fastener
U230	United Steel Deck Inc.	Galvanized steel, galvalume, Al
Ultra-Dek 124	MBCI	Galvalume
Ultra Grip Phillips Head #12	Johns Manville International	Screw – insulation to wood
Ultra Lock 25	IKO Manufacturing Inc.	Organic shingle
Ultra Ply 45 mil	Firestone Building Products	TPO – polyester
Ultra Ply 60 mil	Firestone Building Products	TPO – polyester
Ultra Rubberized Flashing Cement #19	Karnak Corporation	Coating
Ultraclad SBS	GAF Materials Corporation	SBS – glass fiber foil surfaced
Ultrafast/Hex Head	Johns Manville International	Screw fastener
Ultrafast/Hex Head	Johns Manville International	Screw fastener
Ultragard .045 EPDM	Johns Manville International	EPDM – black
Ultragard .045 R EPDM	Johns Manville International	EPDM – black – polyester
Ultragard .060 EPDM	Johns Manville International	EPDM – white
Ultragard .060 FR EPDM	Johns Manville International	EPDM – black
Ultragard .060 R EPDM	Johns Manville International	EPDM – black – polyester
Ultragard SR 60	Johns Manville International	PVC – polyester
Ultragard SR 50	Johns Manville International	PVC – polyester
Ultragard SR 80	Johns Manville International	PVC – polyester
Ultragard SRT-60	Johns Manville International	TPO – polyester
Ultragard SRT-45	Johns Manville International	TPO – polyester
Ultragard V250	Johns Manville International	PVC – polyester
Ultragard V260	Johns Manville International	PVC – polyester
Ultragrip/Phillips Head	Johns Manville International	Screw fastener
Ultraply 0.45	Firestone Building Products	PVC – polyester
Ultraply 0.6	Firestone Building Products	PVC – polyester

Appendix A (Continued)

Trade name	Supplier	Description
Ultra-Shield Built-Up Mastic	GMX Inc.	Asphalt coating
Ultra-Shield Metal Rustproofing	GMX Inc.	Asphalt coating
Ultra-Shield Non-Fibrated Aluminum	GMX Inc.	Asphalt coating
Ultra-Shield White Roof Coating	GMX Inc.	Coating
Ultra-Shield Fibrated Aluminum	GMX Inc.	Asphalt coating
Unibase Acrylic Adhesive/Primer	United Coatings	Acrylic coating
Unibase Primer	United Coatings	Acrylic
Uniline RP	United Steel Deck Inc.	Many metals
Uni-Lok	United Steel Deck Inc.	Many metals
Unirib C336	United Steel Deck Inc.	Many metals
Uniseal	United Coatings	Epoxy coating
Uniseal Epoxy Sealer	United Coatings	Epoxy coating
Unisil	United Coatings	Silicone coating
Uni-Tile Epoxy Sealer	United Coatings	Epoxy coating
Uni-Tile Sealer	United Coatings	Epoxy coating
Urethane 4 7441 Series Top Coat	Neogard	Urethane
Utility Rib	United Steel Deck Inc.	Galvanized steel
Valoise	Huguenot Fenal	Flat clay tile
Vanguard Flat	Monier Lifetile	Flat concrete tile
Vanguard Roll	Monier Lifetile	Barrel mission concrete tile
Vapor-Chan	Tamko Roofing Products, Inc.	Venting Base Sheet
Velvet Roof Coating #107 AF	Karnak Corporation	Asphalt coating
Vent Solution	Johns Manville	Vent Solution
Venting Base	Firestone Building Products	Venting Base
Veral Aluminum	Siplast Inc.	SBS – glass mat & scrim granule surf.
Veral Copper	Siplast Inc.	SBS – glass mat & scrim Al foil
Veral Stainless Steel	Siplast Inc.	SBS – glass mat and scrim copper foil
Vermont & New York Roofing Slate	Hilltop Slate, Inc.	Natural roofing slate
Vermont Roofing Slate	Evergreen Slate Company Inc.	Natural roofing slate
Verona	Westile, Littleton, CO	Barrel mission concrete tile
Versaply 40	Garland Company Inc.	SBS glass fiber sand

Versaply 60	Garland Company Inc.	SBS glass fiber slag
Versaply 80	Garland Company Inc.	SBS glass fiber slag
Versaply Mineral	Garland Company Inc.	SBS glass fiber granule surfaced
Versaweld Premier	Versico Inc.	TPO – polyester
Versaweld Premier EF	Versico Inc.	TPO – polyester
Versigard EPDM .045	Versico Inc.	EPDM – black
Versigard EPDM .060	Versico Inc.	EPDM – black
Versigard I White	Versico Inc.	EPDM – white
Versigard II FR EPDM .060	Versico Inc.	EPDM – black
Versigard II FR Reinforced	Versico Inc.	EPDM – black – reinforced
Versigard III EPDM .045	Versico Inc.	EPDM – black
Versigard PE EPDM .050	Versico Inc.	EPDM – black
Versigard Reinforced EPDM .045	Versico Inc.	EPDM – black – reinforced
Vertical Seam	Metal Sales Manufacturing Company	Standing seam metal panels
Vicor Ultra	W.R. Grace & Co. – Conn.	W.R. Grace & Co. – Conn.
Victorian Shingle	Berridge Manufacturing Company	Metal simulated shingle
Villa	Monier Lifetile	Barrel mission concrete tile
Vitaply	Garland Co.	Asphalt coating
VSR	Butler Roof Division	Galvanized steel, galvalume
Vycor Select	W.R. Grace & Co. – Conn.	W.R. Grace & Co. – Conn.
Wallcote System	Topcoat Inc. (GAF)	Synthetic rubber, acrylic
Warrior Choice	Warrior Roofing Manufacturing Inc.	Asphalt organic felt
Weatherking	Garland Co.	Asphalt coating
Weatherking FR Topcoat	Garland Co.	Asphalt coating
Weatherking Plus	Garland Co.	Asphalt coating
Weatherlock G Granulated Surface	Owens Corning Fiberglas	PMA pressure sensitive
Weatherlock M Mat Faced	Owens Corning Fiberglas	PMA pressure sensitive
Weathermaster ST	Atlas Roofing Corporation	Organic – 3 tab shingle
Weather-Watch	GAF Materials Corporation	PMA pressure sensitive
Wet/Dry Cement	ALCM	Asphalt cement
White Knight	Garland Co.	Urethane
Williamsburg	Ludowici Roof Tile Inc.	Flat clay tile

Appendix A (Continued)

Trade name	Supplier	Description
Winterguard Waterproofiing	Certain Teed Corporation	PMA pressure sensitive
Woodscape 25	Certain Teed Corporation	Glass fiber – lam. rand. tab shingle
Woodscape 30	Certain Teed Corporation	Glass fiber – lam. rand. tab shingle
Woodscape 40	Certain Teed Corporation	Glass fiber – lam. rand. tab shingle
XHD	Olympic Manufacturing Group/N.B.T.	Screw fastener
XL Americana	Ludowici Roof Tile Inc.	Flat clay tile
XL Classic	Ludowici Roof Tile Inc.	Flat clay tile
XL Lania	Ludowici Roof Tile Inc.	Flat clay tile
XL Williamsburg	Ludowici Roof Tile Inc.	Flat clay tile
XT 25	Certain Teed Corporation	Glass fiber – 3 tab shingle – in./lb
XT 30	Certain Teed Corporation	Glass fiber – 3 tab shingle – in./lb
Znanchor Drive Nail EGS-PIN	Simplex	Sleeve expansion fastener
Znanchor Drive Nail SS-PIN	Simplex	Sleeve expansion fastener

Appendix B – Roofing industry websites

Organization	Website
(APA) Engineered Wood Association	www.engineeredwood.org
AEP-Span	www.aep-span.com
Air Vent Inc.	www.airvent.com
ALCO-NVC, Inc.	www.alconvc.com
American Concrete Institute	www.aci-int.org
American Institute of Architects	www.aia.org
American Institute of Steel Constructors	www.aisc.org
American Iron and Steel Institute	www.steel.org
American National Standards Institute	www.ansi.org
American Society of Civil Engineers	www.asce.org
American Wood Preservers Institute	www.awpi.org
AMF R-Control Building Systems	www.r-control.com
Architect's Catalogue, Inc.	www.arcat.com
Architects First Source Exchange	www.firstsourceexchange.com
Asphalt Roofing Manufacturers Association	www.asphaltroofing.org
ASTM International	www.astm.org
Atlas Roofing Company	www.atlasroofing.com
Bakor, div. of Henry Corporation	www.bakor.com
Barrett Company	www.barrettroofs.com
Benchmark Foam Inc.	www.benchmarkfoam.com
Berridge Manufacturing Company	www.berridge.com
Bitec Inc.	www.bi-tec.com
Bondcote Roofing Systems	www.bondcote.com
Building Industry Exchange	www.building.org
Building Official Congress America	www.bocai.org
Buildings Magazine	www.buildings.com
Canadian Codes Centre	www.codes.nrc.ca
Canadian Standards Association	www.csa.ca
Carlisle Syntec Inc.	www.carlislesyntec.com
Cedar Shake & Shingle Bureau	www.cedarbureau.org
Certain Teed Corporation	www.certainteed.com
Conklin Company	www.conklin.com
Construction Fasteners Inc.	www.constructionfasteners.com
Construction Specifications Institute	www.csinet.org

Appendix B (Continued)

Organization	Website
Cool Roofs	http://eetd.lbl.gov/heatisland
Cooley Engineered Membrane Inc.	www.cooleygroup.com
Copper Development Association	www.copper.org
Corps of Engineering Research Labs	www.cecer.army.mil/homepage. html
Corps of Engineers Cold Region Lab	www.crrel.usace.army.mil
Curveline Inc.	www.met.tile/com/curveline
Danosa Caribbean Inc.	www.danosapr.com
Dewitt Products Company	www.dewitt@globalbiz.com
Duro-Last Inc.	www.duro-last.com
Ecology Roof Systems	www.ecologyroofsystems.com
Elastomeric Roof Systems Inc.	www.ersystems.com
Elk Corporation	www.elkcorp.com
Energystar Program	www.energystar.gov
Eternit, Inc.	www.eternitusa.com
Evergreen Slate Company	www.evergreenslate.com
Facility Management	www.facilitymanagement.com
Factory Mutual Research Laboratory	www.fmglobal.com/index.html
FEMA Federal Emergency Management Agency	www.fema.gov
Firstsource	http://firstsource.com
GAF Materials Corp.	www.gaf.com
Genflex Roofing Systems	www.genflex.com
Green Building Council	http://usgbc.org
Gypsum Association	http://gypsum.org
Henry Company	www.henry.com
Hilltop Slate Inc.	www.hilltopslate.com
Huebert Fiberboard Co.	www.huebertfiberboard.com
ICBO	www.icbo.org/index.html
IKO Industries	www.iko.com
International Code Council	www.intlcode.org
IPS Insulated Panel Systems	www.insulated-panels.com
Karnak Corporation	www.karnakcorp.com
Koppers Industries Inc.	www.koppers.com
Laurence Berkeley National Laboratory	http://eandE.LBL.gov/heatisland
Links on Roofing	www.rooftex.com/links.html
M.C.A. Clay Roof Tile	www.mca-tile.com
Malarkey Roofing Company	www.malarkey-rfg.com
MBCI	www.mbci.com
Metacrylics Acrylic	www.metacrylic.com
Metal Building Manufacturers Association	www.mbma.com
Metal Construction Association/Metal Roofing Alliance	www.metalconstruction.org
Metal Sales Manufacturing Company	www.milsales.com
Met-Tile Inc.	www.met-tile.com/roof
Midwest Roofing Contractors Association	www.MRCA.org
Monier Lifetile	www.monier.com

National Coil Coating Association	www.coilcoating.org
National Concrete Masonry Association	http://ncma.org
National Fire Protection Association	www.wpi.edu%7Efe/nfpa.html
National Research Council – Canada	www.nrc.ca.irc/roofing
National Roofing Contractors Association	www.nrca.net
Nonwoven Fabric Institute	www.inda.org
North American Insulation Manufacturers Association	www.naima.org
North Carolina Foam Industries	www.ncfi.com
OSHA	www.osha.gov
Owens Corning	www.owenscorning.com
Pacemaker Plastics Company	www.pacemakerplastics.com
Pact IV Building Products	www.pactivbuildingproducts.com
Peterson Aluminum	www.pac-clad.com
PIMA	www.pima.org
Pittsburgh Corning Corporation	www.foamglasinsulation.com
Plymouth Foam Inc.	www.plymouthfoam.com
Polyglass USA	www.polyglass.com
Precast Prestressed Concrete Institute	www.pic.org
Rilem	http://web.ens-cachan.fr.80/~rilem
R-Max Inc.	www.minc.com
Roof Coating Manufacturers Association	www.roofcoating.org
Roof Consultants Institute	www.RCI-online.org
Roofing Contractor Magazine	www.roofingcontractor.com
Roofing Industry Educational Institute	www.riei.org
Roofing Products International	www.roofingproductsint.com
Roofing Technology Favorite Sites	www.roofingtech.com/links.html
RSI Magazine	www.RSIMag.com
Sarnafil, Inc.	www.sarnafilus.com
Scientific and Technical Information Network	http://info.cas.org/stn.html
Seaman Corporation	www.fibertite.com
Senco Products	www.senco.com
Simpson Gumpertz & Heger Inc.	www.sgh.com
SMACNA	www.smacna.org
Somay Products Inc.	www.somay.com
Southern Building Code Conference	www.sbcci.org
Southwestern Petroleum Company	www.swepcousa.com
Spray Polyurethane Foam Alliance	www.sprayfoam.org
SPRI Sheet Membrane & Component Suppliers	www.spri.org
Steel Deck Institute	www.sdi.org
Steel Roofing (Galvalume)	www.steelroofing.com
SWD Urethane Company	www.swdurethane.com
Sweets Catalogue	www.sweets.com
System Builders Association/Metal Buldings Institute	www.systemsbuilders.org
Tamark Manufacturing, LLC	www.lifepine.com
Tamko Roofing Products	www.tamko.com
Temple	www.temple.com
Texas Refining Corp.	www.texasrefinery.com

Appendix B (Continued)

Organization	Website
Thomas Register	www.thomasregister.com
Tremco Inc.	www.tremcoroofing.com
True-Fast Corporation	www.trufast.com
U.S. Intec Inc.	www.usintec.com
United Coatings	www.unitedcoatings.com
US Army Corps of Engineers, technical	www.hnd.usae.army.mil/techinfo
Vincent Metal Goods	www.vincentmetalgoods.com
W.P. Hickman Systems, Inc.	www.wphickman.com
W.R. Grace & Co. – Conn.	www.gcp-grace.com
Western Roofing Magazine	www.westernroofing.net
Western States Roofing Contractors Association	www.wsrca.com
Westile	www.westile.com

Appendix C – Answers

PART I

Chapter 1   1 a; 2 a; 3 a; 4 d; 5 a; 6 c; 7 b; 8 b; 9 a; 10 a; 11 a; 12 d.

Chapter 2   1 a; 2 b; 3 b; 4 c; 5 a; 6 b; 7 c; 8 b; 9 a; 10 b; 11 a; 12 b; 13 b; 14 d; 15 b; 16 a; 17 a; 18 b; 19 a; 20 a; 21 b; 22 b; 23 b; 24 a; 25 a; 26 a; 27 b; 28 c; 29 b; 30 a; 31 a; 32 a; 33 b; 34 a; 35 b; 36 a; 37 a; 38 b.

Chapter 3   1 d; 2 b; 3 a; 4 b; 5 a; 6 b; 7 b; 8 a; 9 c; 10 a; 11 c; 12 a; 13 b; 14 b; 15 a; 16 a; 17 b; 18 b; 19 b; 20 b; 21 a; 22 a; 23 b; 24 b; 25 a; 26 b; 27 a; 28 b; 29 a; 30 a.

Chapter 4   1 a; 2 b; 3 a; 4 a; 5 b; 6 b; 7 b; 8 a; 9 b; 10 a; 11 a; 12 b; 13 a; 14 a; 15 a; 16 a; 17 a; 18 b; 19 b; 20 b; 21 a; 22 a; 23 a; 24 a; 25 b.

Chapter 5   1 a; 2 c; 3 b; 4 a; 5 d; 6 b; 7 a; 8 d; 9 a; 10 a; 11 b; 12 b; 13 b; 14 a; 15 b; 16 a; 17 b; 18 b; 19 b; 20 b; 21 a; 22 a; 23 b; 24 b.

Chapter 6   1 a; 2 b; 3 b; 4 a; 5 e; 6 c; 7 a; 8 b; 9 b; 10 b; 11 a; 12 a; 13 c; 14 b; 15 b; 16 b; 17 a; 18 b; 19 a; 20 a; 21 a; 22 a; 23 d; 24 b; 25 a.

Chapter 7   1 a; 2 a; 3 b; 4 a; 5 b; 6 a; 7 a; 8 a; 9 b; 10 b; 11 b; 12 a; 13 b; 14 a; 15 b; 16 a; 17 a; 18 b; 19 a; 20 a.

Chapter 8   1 a; 2 a; 3 b; 4 b; 5 a; 6 b; 7 b; 8 a; 9 a; 10 a.

Chapter 9   Probably the lack of effective supervision or monitoring was the chief cause of failure. Many of the conditions such as the "M" shaped deck should have been corrected before the roofing was installed. A settlement was reached in this case. As with many settlements, the details are confidential. An effective monitor could have prevented this failure.

Chapter 10   The principal cause of failure was the marketing of a product that was unsuitable for use. Avoid products that only have a short performance history. Here the designer was at fault, but he relied on the special expertise of the manufacturer. That reliance was misplaced. The owner and the manufacturer reached a private settlement.

Chapter 11   The original splits were caused by the use of glass foam insulation on a steel deck of inadequate stiffness. This is against the insulation manufacturer's recommendations. Under load the deck deflected between the joists so that the joints between the insulation panels over the joists rotated open to tear the membrane. The fix merely reduced the number of rotating open joints to those where the insulation and gypsum board butt joints were aligned over or near a joist. Lesson: Avoid planes of weakness; they become stress concentrators. Careful planning of the layout of materials might avoid this kind of problem.

Chapter 12   There seems to be enough blame here to spread it over all the participants including the owner, designer, installer, and material suppliers.

- The Government should require the designer to select the materials to be used; that would make the responsibility clear.
- Designers should avoid new products unless there is an unusual need for the product and the owner is made fully aware of the experiment.
- Roofers should avoid installing products that they have not previously installed without advising all concerned of their lack of experience.
- Manufacturers should not sell products that have not been tested in the field.

Chapter 13   The fault here is difficult to assign. I suggest that the gypsum supplier should have had more detailed instructions to stagger the butt joints of the bulb tees on alternate bar joists, and to require lapping a wire tying the reinforcing wire sheets. This might have been picked up in a peer review.

Chapter 14   Disagreements with your client are always difficult. Greater problems may appear if you bow to irrationality, but failure to document your good advice may not serve you or your client in the long run. Do your best to clear up any misunderstandings, and try and protect your client from his own lack of knowledge or experience.

Chapter 15   The jury found that the fire was caused by an electrical short circuit (an Act of God). This protected the locals who attempted to install the roof without experience or insurance. Being sure that the experience was present could probably have prevented this fire, to say nothing of the insurance, which a good roofing contractor could have provided.

Chapter 16   The water as a result of condensation from the humidification system used. The parapets at the perimeter of the building were acting as chimneys. The excessive air pressure within the building forced the humidified air up the inside of the parapets; the moisture condensed on the cool surfaces and drained down into the building. The design work was at fault. Be very careful of having a significantly different air pressure inside and outside a building. In northern climes, a positive interior pressure can force moist air out to the cold building envelope, to condense and do evil things. In southern climes, a negative internal pressure can pull moist air into and onto colder air conditioned spaces, to condense and promote fungus and rot. Most difficult are vacation homes or buildings with seasonal occupancy, because ventilation variation can result in problems anywhere.

Chapter 17   The ignition source is unknown to this day. The case was eventually settled with the alarm company paying the largest share of the settlement. The roofing materials manufacturer made a modest payment.

Chapter 18 Yes, failure did occur. The product was not suitable for the purpose. The manufacturer may be guilty of fraud, since the manufacturer knew or should have known that the product could not last for the warranty period. This type of event is currently the subject of several class action suits brought by dissatisfied building owners.

Chapter 19 Poor design is the basic problem. You must separate roofing areas supported by different structures with an expansion joint. You must not rely on drainage across an expansion joint. You should provide expansion joints wherever movement can be expected – here at all the perimeters. The contractor was not responsible for the leakage. Failure to provide a means for maintenance access is also a design defect.

Chapter 20 The supplier is at fault. Pitch and asphalt should not be combined. I doubt that a warning would help, because the supplier was in a state of denial. The roofer had no responsibility. Total removal is the only reasonable solution – paid for by the supplier. This is my opinion – I don't know what really happened.

Chapter 21 I am appalled by the actions of the engineer and lawyer. This kind of nonsense should be fought to a standstill in court. Unfortunately, few cases involve enough money to justify the legal proceedings. This case was settled with a modest payment from the supplier.

Chapter 22 Part of the fault goes to each party except the owner. The designer should not have specified or approved the use of phenolic foam insulation. The general contractor should have had better control of the work so the built-up roof was not used as a work platform. The roofer should have provided better protection during the construction of the mansards. Of course, the supplier should have never marketed the phenolic foam.

Chapter 23 The designer provided a specification that was not clear. In addition, the supplier provided an experimental product that had no history of excellent performance. This type of problem can be avoided by using time-tested products from reputable manufacturers. On large jobs pretest the shingles to be sure they meet the specifications. These shingles have to be replaced.

Chapter 24 Insulation shrinkage is the major mechanism for the splits. Splitting is more likely because of stress concentrations over the insulation joints. Peer review, or the use of a cover board would probably have avoided this problem. At this time no major roofing material manufacturer would warrant their products over this insulation. The designer should have been warned against its use.

Chapter 25 Attempting to grow or maintain market position by making products that does not even meet the minimum ASTM standards just does not make economic sense. The perpetrator is likely to lose his credibility in the industry, and is likely to have to make good for all the roofs that do not properly perform at a higher cost than the money saved.

Chapter 26 The design did not take into account the water that is the natural part of the lightweight-insulating concrete installation. It is probable that the designer relied on the special expertise of the supplier, but either it was not consulted or improper advice was given. Like almost every material, lightweight-insulating concrete can work if it is properly used. Total removal is necessary on this job. Peer review should have prevented this failure.

Chapter 27 One of the many problems here was that the supplier provided green insulation that had been hurried through the manufacturing process without a reasonable curing time to attain a stable size. Part of the problem may be the higher interior pressure forcing moist air outward to cooler surfaces, and perhaps peeling open the EPDM joints before they developed their full strength. An effective air barrier is a must whenever you have significant pressure differentials across a system. The specified construction was suitable for most of the year, but an additional air and vapor barrier was needed during the winter months to prevent the moist interior air from contacting the cold underside of the membrane. The adhesion between roofing components was poor to marginal. We are uncertain if this is due to the quantity of the adhesive used by the contractor. The contract specifications were not very clear on the subject. Peer review may have caught many of these problems, and monitoring reports might have answered some of the questions we have about the application of the materials.

Chapter 28 This is another one of the asphalt and coal-tar pitch incompatibility cases. In this beauty, the pitch supplier claimed that the asphalt-glass felt supplier gave him poor or misleading information. The roofing materials suppliers were at fault because the system could not work even if angels had installed it. The school board was correct. Remove and replace the entire system.

Chapter 29 Freezer's are so prone to problems that often special crews who appreciate the problems involved are used in their construction. When a contractor is hired in these difficult cases, the owner is not, or should not be hiring, just a labor broker. He should be dealing with craftsmen knowledgeable in the field. Here it was evident that the contractors did not have the special skill that is appropriate. I don't know if the owner, the contractor, or the job supervision were more to blame.

Chapter 30 Unreasonable claims will probably exist as long as we have lawyers. One way to get a reasonable settlement is to make an unreasonable demand (twice what you want), then settle for half of your demand. Litigation and condominiums are closely linked. You should expect a suit as soon as the ownership moves from the developer to the condo association. There are lawyers who specialize in that type of litigation. They might even have check list forms listing typical complaints. I suspect some bring suit using a laundry list they never checked. Some lawyers expect the developers to quickly settle for some relatively small amount without even checking the validity of the claim. Trust everyone – but check.

Chapter 31 The solution to this problem eluded us for a long time, and I'm not even sure we have the right answer now. As usual, I suspect that many elements had to be involved including the lack of slope to drains, and the weak asbestos felts. This was a singular occurrence. I hope it is not repeated elsewhere.

Chapter 32 Somehow the coated felt plies got wet before they were installed on the roof. It is also probable that the base (the bottom) ply was installed some time before the top ply was installed. You will see that moisture tends to gather and dry out last at the edge of the coated felt lap. Even the small quantity of water associated with dew is dangerous; it is bound to create blistering. In this case the materials supplier and the roofer replaced the roof at their cost – we never found out how the costs were split. Filling the blisters with mastic is a good idea on paper, but it is not practical in the field, where blisters are not uniform and have holes in varying degrees.

Chapter 33 In the case of the Canadian EPDM roof, the fasteners were not suitable for the purpose. Since the EPDM manufacturer supplied them, the manufacturer was liable. This was at a point in time when the industry was just learning about the dangers of wind. Since then much more effective fasteners have been developed that incorporate all kinds of features to prevent fastener backout. As discussed elsewhere, fasteners should not be installed near the membrane. Use the fasteners to attach the bottom layer of insulation, then use a cover board or another adhered insulation layer to cover the heads of the fastener before the membrane is adhered.

Chapter 34   This blistering in an asphalt-glass fiber shingle is a rare event. It is so rare that testing for a shingle's blistering tendency is probably not worthwhile. This case is reported so that remedial work can quickly be performed whenever shingle blistering is observed. Blistering in asphalt-glass fiber shingles is a product defect. The supplier should make the best deal he can with the roofer and get the shingles replaced.

Chapter 35   Aside from the point that bad news travels fast and far, there is a real occasional danger that greed or fear causes owners to take positions that cannot be supported by a reasonable investigation. It is hard to prevent this because their position is based on emotions rather than facts. Get the facts before launching a lawyer-powered dispute. Hire a competent investigator that will give you a realistic report on the condition of your roofing system. If you are retained to investigate a problem, investigate sufficiently to allay the fears of the owner; be sure he received the value for which he is entitled.

Chapter 36   The suspicion here is that the owner or purchasing agent thought they were getting a new roof at a bargain price. After all, they could avoid using those expensive designers and roofing contractors – not to mention the cost of a design peer review, or a monitor for the roofing work. The supplier's salesman said he would supervise the installation personally! (He never did appear after the first day.) The temptation is great to say the owner got what he deserved, but that accurate and unkind thought does not consider that he did not get what he bargained for. He was entitled to a roof that did not leak. The supplier ultimately paid for the removal of his mess and returned the funds paid for the new roof.

Chapter 37   The fundamental point of this case is that any pressure sensitive adhesive must be dry, cure, or otherwise become hard if it is going to survive exposure to the weather. That is why duct tape is not suitable for use as a permanent roofing material, contrary to many contractors' apparent expectations. Sheets with pressure sensitive asphalt are currently used as shingle underlayment and waterproofing; these are anchored in place by other materials. They cannot perform properly when they are exposed to the weather for any substantial interval. Again, peer review should avoid this problem.

Chapter 38   There is no question that the shingles have to be removed to repair the plywood deck over the party walls. After ten years of exposure there is no way the shingles can be matched, so all the shingles have to be removed. There is also no doubt that the shingles failed after ten years of service, but new shingles could have been installed over the torn shingles if the plywood deck was in good condition. In this case, the good faith agreement between the supplier of the plywood and the supplier of the shingles to apportion the cost to replace the offending materials is needed. I wonder if their insurers will permit an agreement without a law suit?

Chapter 39   Here we are dealing with new technology. After discovering the apparently poor dispersion in the product that did not work, we checked the dispersion of candidate replacement products for the client before they were installed – rejecting several candidates. The technology is so new and there is no standard in place; there is a danger of rejecting a material that will perform adequately. At the moment, the asphalt dispersion in polymer should be reported to be compared to performance until a much larger database is assembled.

Chapter 40   The first problem was to provide drainage at the waterproofing layer – however poor was the quality of the membrane. We had the plumber remove the drain fittings, and watched the water pour into the drains. Liquid applied waterproofing without reinforcement is damp-proofing, not waterproofing, but that is what was specified. The dead trees in the planters all had drowned because the planters were not drained. The fountain leaked because of poor design combined with sloppy workmanship permitted by inadequate supervision. Designers of mechanical equipment such as fountains and air conditioning equipment often pay little attention to how their equipment is going to be installed or the linkage between their wonderfully designed unit and the balance of the building or structure, or the effect of equipment operation on the surrounding materials. In this case, the massive stainless steel water intake was a thing of beauty, but there was no way to attach it to the structure nor and way to moderate the vibrations when the fountains pump was in operation. Peer review would have caught these problems. Monitoring would have handled the defective workmanship.

Chapter 41   Roofing membranes are seldom attached directly to structural concrete decks. This lack of experience was surely a factor here. Normally insulation acts as a thermal barrier, an attenuating layer, and a layer to horizontally distribute minor local air and moisture pressures. These latter functions are not often understood or appreciated. In this case the supplier was not aware of the potential problem, but remediated the situation as soon as it was understood.

Chapter 42   The sea gulls aside – the problem here was inadequate application of adhesive. Effective monitoring would have prevented this. There are times when monitors spent their time in their car and not on the roof. On other occasions, one member of the roofing crew was detailed the job to keep the monitor busy away from the roofing work. These are just two examples of wasted monitoring. Your monitor must be knowledgeable and on top of the work. One of our monitors told a roofing contractor that he had to turn in his keel (the crayon used to mark defects) each night to be weighed – to be sure he it often enough – and he was believed!

Chapter 43   The fundamental problem here was the excessive distance between expansion joints or the perimeters of the roof. Fixity was provided at the perimeter. The central ridge provided a slight degree of stabilization. Each roof was large enough to cover two American football fields, fastened to the steel deck with mechanical fasteners that prevented uplift, but did little to stay horizontal movement as evidenced by the ovated holes in the insulation. Note that the only roof area that did not split contained a large number of skylights; each providing the fixity needed to stabilize the membrane. This design was defective. A peer review would probably have been picked up the potential problems.

Chapter 44   This case demonstrates that the vacuum box "performance test" is relatively useless in predicting performance in a wind. The forces of the winds actually experienced were much lower than the forces measured by the test, yet the roof failed. Aside from reliance on an inadequate test method, optimistic workmanship was the cause of this failure. They apparently felt they would obtain enough adhesion by changing the method of adhesive application (they always passed the vacuum box test), but it was not good enough. It is beneficial to apply a primer coating to increase adhesive hold-out in adhesive applications to low density materials. This would improve adhesion.

Chapter 45   Bentonite is not recommended for horizontal applications nor applications where it clay is not confined. Wet and dry cycles can also harm bentonite waterproofing because the shrinking drying jell will crack and not reseal upon rewetting. This was a misuse of a marginal material.

Chapter 46   Poor venting is not a valid excuse for poor shingle performance. Improperly designed shingles causes thermal splits. It would seem quite dangerous to replace improperly designed shingles with the same improperly designed shingles. For every important job, have the shingles analyzed for tear strength before they are installed.

Chapter 47   Maybe we were at fault for not insisting on a cover board, but the real enemies are those whose main desire is lower-in-cost rather than greater-in-value. Good roofing is not simple or easy. The roofing system needs all the help it can get to survive in this world bent on its destruction.

Chapter 48   Reducing the temperature fluctuations seen by the system can solve this problem. The acrylic coating might work, but would probably have to be replaced every three years. A white Hypalon coating might last longer, but we are not sure that the temperature variation reduction would be enough to quiet the system. An alternate suggestion is to slash the existing EPDM membrane, install another layer of insulation with mechanical fasteners, and a new EPDM membrane. The added insulation would moderate the temperature swings and stiffen the system. The fault here is with the designer for failing to follow the manufacturers recommendations. Peer review would probably prevent this problem – if the designer listened to the reviewer.

Chapter 49   The roof on the elementary school needs to be replaced – because of age, not hail damage. The high school needs some flashing repairs, and the glass in the conservatory or green house should be replaced due to the hailstorm. There is no action required on the junior high school roof. You should design and build to resist hail in hail prone areas. This means a dense insulation under the membrane, a strong membrane, and a surface protected with gravel or pavers.

Appendix D – Bibliography

MAJOR REFERENCE WORKS

Baker, M.C. (1980) *Roofs – Design, Application and Maintenance*, Ottawa: National Research Council of Canada

Griffin, C.W. and Fricklas, R. (1996) *The Manual of Low-Slope Roof Systems*, New York: McGraw-Hill

Henshell, J. (2000) *The Manual of Below – Grade Waterproofing Systems*, New York: John Wiley & Sons, Inc.

Hoiberg, A.J. (1963) *Bituminous Materials: Asphalts, Tars, and Pitches*, New York: Interscience Publishers, John Wiley & Sons

Laaly, H.O. (1993) *The Science and Technology of Traditional and Modern Roofing Systems*, Glendale, CA

NRCA (2001) *The NRCA Roofing and Waterproofing Manual*, 5th edn, Rosemont, IL: National Roofing Contractors Association

NRCA (biannually) *Low-Sloped Roofing Materials Guide*, Rosemont, IL: National Roofing Contractors Association

NRCA (biannually) *Steep-Sloped Roofing Materials Guide*, Rosemont, IL: National Roofing Contractors Association

Revere Copper Products, Inc. (2000) *Copper and Common Sense*, 7th edn, Rome, NY, USA

SMACNA (1993) *Architectural Sheet Metal Manual*, Chantilly, VA: Sheet Metal and Air Conditioning Contractors National Association

OTHER DOCUMENTS OF INTEREST

Anderson, L. (1989) *Building Failures*, London: Butterworth Architecture

ARMA (2000) *Residential Asphalt Roofing Manual*, Lexington, KY: Asphalt Roofing Manufacturers Association

Cash, C.G. (1994) 'Estimating the Mean Temperature of Horizontal Surfaces for Predicting the Durability of Thermally Sensitive Materials', *Dealing with Defects in Buildings*, Varenna, Italy: CIB-W86

Cash, C.G. (2000) 'Estimating the Durability of Roofing Systems', *Durability 2000, Accelerated and Outdoor Weather Testing*, West Conshohocken, PA: ASTM STP 1395

Cash, C.G. (2000) 'Roofing Materials', *The Encyclopedia of Materials: Science and Technology*, Chapter 11, London: Elsevier Science Ltd

Hardy, S. (1998) *Time-Saver Details for Roof Design*, New York, NY: McGraw Hill

Kaminetzky, D. (1991) *Design and Construction Failures*, New York: McGraw-Hill
Property Services Agency (1989) *Defects in Buildings*, London: Her Majesty's Stationery Office
Ransom, W.H. (1987) *Building Failures*, 2nd edn, London: E. & F.N. Spon
Scharff, R. *et al.* (1996) *Roofing Handbook*, New York: McGraw-Hill

CITED WORKS

Abraham, H. (1945) *Asphalts and Allied Substances*, New York: D. Van Nostrand Co., Inc.
Loss, J. (February 1987) *Update*, University of Maryland: Architecture and Engineering Performance and Information Center
Bailey, D.M., Cash, C.G. and Davies, A.G. (2002) *Predictive Service Life Tests for Roofing Membranes*, Champaign, IL: Construction Engineering Research Laboratory, ERDC CERL TR-02-22
Cash, C.G. (1998) *The Durability of Roofing Systems*, Presentation: Institute of Roofing and Waterproofing Consultants, Las Vegas, NV
CIB-W86 (1993) *Building Pathology a State-of-the-art Report*, Publication 155
Green, P. (February 1987) 'Roofing: Synthesizing Design and Craftsmanship', *Architectural Record*, New York: McGraw Hill
Hinojosa, O. and Kane, K. (April 2002) 'A Measure of the Industry', *Professional Roofing*, Rosemont, IL: NRCA
Mathey, R.G. and Cullen, W.C. (1974) *Preliminary Performance Criteria for Bituminous Membrane Roofing*, Washington, DC: National Bureau of Standards
Munich Reinsurance Company (1984) *Munchener Ruck*, D-8970 Immenstadt: Eberl GmbH
Alberti, L.B. (1695) *10 Books on Architecture*

ASTM INTERNATIONAL PUBLICATIONS

ASTM (1999) *ASTM Standards Relating to Materials, Systems and Testing for Roofing and Waterproofing*
ASTM (2001) *Annual Book of ASTM Standards Vol. 04.04 on Roofing and Waterproofing*
C1289 – *Standard Specification for Polyisocyanurate Thermal Insulation Board*
C208 – *Standard Specification for Cellulosic Fiber Insulating Board*
C407 – *Standard Specification for Roofing Slate*
C552 – *Standard Specification for Cellular Glass Thermal Insulation*
C578 – *Standard Specification for Preformed, Cellular Polystyrene Thermal Insulation*
C726 – *Standard Specification for Mineral Fiber Roof Insulation Board*
C728 – *Standard Specification for Perlite Thermal Insulation Board*
D1863 – *Standard Specification for Mineral Aggregate Used on Built-Up Roofs*
D3018 – *Standard Specification for Class A Asphalt Shingles Surfaced With Mineral Granules*
D312 – *Standard Specification for Asphalt Used in Roofing*
D3161 – *Standard Test Method for Wind Resistance of Asphalt Shingles (Fan Induced Method)*
D3462 – *Standard Specification for Asphalt Shingles Made from Glass Felt and Surfaced with Mineral Granules*
D36 – *Standard Test Method for Softening Point of Bitumen (Ring-and-Ball Apparatus)*

D4073 – *Standard Test Method for the Tensile-Tear Strength of Roofing Membranes*

D4402 – *Standard Method of Test for Viscosity Determinations of Unfilled Asphalts Using the Brookfield Thermocel Apparatus*

D4932 – *Standard Test Method for Fastener Rupture and Tear Resistance of Roofing and Waterproofing Sheets*

D4989 – *Standard Test Method for the Apparent Viscosity (Flow) of Roofing Bitumens Using the Parallel Plate Plastometer*

D5385 – *Standard Test Method for Hydrostatic Pressure Resistance of Waterproofing Membranes*

D5601 – *Standard Test Method for Tearing Resistance of Roofing and Waterproofing Membranes*

D5602 – *Standard Test Method for Static Puncture Resistance of Roofing Membrane Specimens*

D5635 – *Standard Test Method for Dynamic Puncture Resistance of Roofing Membrane Specimens*

D92 – *Standard Test Method for Flash and Fire Points of by Cleveland Open Cup*

E632 – *Standard Practice for Developing Accelerated Tests to Aid in the Prediction of the Service Life of Building Components and Materials*

FACTORY MUTUAL RESEARCH CORPORATION DOCUMENTS

(Annual), Factory Mutual Research Approval Guide

1–28 (June 1996) 'Wind Loads to Roofing Systems and Roof Dck Securement', *Property Loss Prevention Data Sheets*

1–29 (June 1996) 'Above Deck Components', *Property Loss Prevention Data Sheets*

1–29 (October 1984) 'Adhered or Mechanically Attached Single-Ply Membrane Systems', *Property Loss Prevention Data Sheets*

1–29 attachment (October 1984) 'Loose-Laid Ballasted Roof Coverings', *Property Loss Prevention Data Sheets*

1–49 (June 1985) 'Perimeter Flashing', *Property Loss Prevention Data Sheets*

1–60 (2000) 'Asphalt Coated Metal and Protected Metal Buildings', *Property Loss Prevention Data Sheets*

Index

Note: Page numbers given in *italics* refer to tables and figures.

adhesion
  asphalt-glass fiber built-up
    membranes 28
  attachment methods 35
  case studies 161–2, 176–7, 182–3
  missing facer adhesion 182–3
  pressure sensitive 161–2
  seal-tab 56–7
  wind problems 176–7
aluminium 60
APP (atactic polypropylene) 6, 29
architectural metal systems 8, 34–5
  copper 59–60
  galvanized/zinc coated steel 60
  lead 60
asbestos 90–1
  case study 148–9
asphalt and coal-tar pitch
  case study 142–3
  flash point 22–3
  grades 22
  historical use 22
  problems 24
  production 23–4
  sources 21–2
  viscosity-temperature
    relationship 22–3
asphalt dispersion (case study) 165–6
asphalt shingles 52–9
  case studies 122–3, 136–7, 154–6
  felts 53–4, 57, 58
  fillers 58–9
  flexible 59
  granule surfacing 59
  individual 52
  laminated 59
  metric 53

  popularity 52
  seal-tab feature 55–7, 58
  tear strengths 58
  thermal splitting 56–7
  thick butt 54–5
  three-tab 52
asphalt-glass fiber built-up membranes
  adherence 28
  back nailing 28
  composition 27
  precautions needed 28
  suitability/durability 27
asphalt-glass fiber shingles 7–8
attachment methods
  adhered membranes 35
  ballasted systems 35
  mechanical 35–6

beadboard *see* EPS (expanded
  polystyrene)
bentonite 178–9
blistering (case studies)
  airport roof 169–70
  phased felt plies 150–1
  shingles 154–6
built-up roofs *see* asphalt-glass fiber
  built-up membranes

case studies
  artificial slates 105–6
  asbestos 148–9
  asphalt dispersion 165–6
  blistered shingles 154–6
  cast-in-place concrete decks 124–5
  ceramic tiles 146–7
  cold process roofing 159–60
  condominiums 128–9

case studies (*Continued*)
  corroding foam 130–1
  dissolving shakes 122–3
  distant expansion joints 173–5
  DIY/maintenance failure 115–16
  fastener blackouts 152–3
  fire retardant plywood 163–4
  flapping glass fiber felts 126–7
  gooey felts 142–3
  hail storm 186–7
  heavy glass fiber shingles (case
      study) 132–3
  improper waterproofing 178–9
  leaks 101–4, 110, 117–18, 124–5,
      128–9, 144–5
  lightweight insulating concrete
      138–9
  liquid applied waterproofing 167–8
  missing facer adhesion 182–3
  moving insulation 134–5
  noisy roof 184–5
  phantom deck movement 136–7
  poor ventilation 180–1
  pressure sensitive adhesion 161–2
  propriety products 109–10
  sea gulls 171–2
  shingle splits 136–7
  shrinking insulation 140–1
  skaters' cracks 148–9
  split membranes 107–8, 111–12
  two ply problem 157–8
  unacceptable design advice 113–14
  vermiculate concrete 138–9
  widespread flame 119–20
  wind effect on adhesion 176–7
ceramic tile 63
  case study 146–7
coal-tar *see* asphalt and coal-tar pitch
cold process system (case study)
    159–60
concrete
  lightweight insulating 41, 138–9
  lightweight structural 40
  poured-in-place 40
  precast/pre-stress panels 42
  tile 63
  vermiculate case study 138–9
condominiums (case studies) 128–9, 133
cool roofing 96
corroding foam (case study) 130–1
corrugated metal 60–1
CPE (chlorinated polyethylene) 33
CSPE (chlorosulfanated
    polyethylene) 33

decks *see* structural decks
design advice (case study) 113–14
DIY/maintenance failure (case
    study) 115–16
drainage 19–21
dual-slope systems 5

Environmental Protection Agency 58
EPDM (ethylene-propylene-diene
    terpolymer) 5, 7, 30–1, 44, 61, 96
  case studies 140–1, 171–2, 184–5
EPS (expanded polystyrene) 47
expansion joint 77, 78, 79
  case study 173–5

Factory Mutual Research Corporation
    (FMRC) 38
failure
  asbestos 90–1
  case studies 99, 101–90
  contractor responsibility 88
  defective design 87, 88, 89
  defective workmanship 89
  definitions 84–5
  evidence 85
  examples 85
  identifying 86–7
  innovation 91–2
  lack of product information 91
  laundry lists of errors 86
  life-span expectations 92
  materialmen/manufacturers
      contribution 89–92
  owner responsibility 87
  product promotion/marketing
      myopia 90
  reasons 87
  testing of systems 90
  warranties 89
fasteners (case study) 152–3
felts 53–4, 57, 58
  asbestos 148–9
  case studies 142–3, 148–9, 150–1,
      157–8
  gooey 142–3
  organic 54, 57
  phased felt plies 150–1
  rag 53–4
  shingled 158
fiber-cement tiles/slates 63–4
fillers 58–9
fire
  case studies 119–20, 163–4
  classification 52, 53, 57–8

fire retardant plywood 163–4
Standard (ASTM D92) 22
Standard (ASTM D3018) 94
wide spread flame 119–20
flashing
  built-up/polymer modified
      bitumen 69
  consultation with designer 68–9
  coping caps 79–80
  counterflashing 80, *81*
  curb detail 71, *79*
  errors/failures 66–7
  expansion joint 77, *78, 79*
  face reglets 67
  fastener patterns 79
  gravel stops/metal edge 75, *76*
  incomplete cap flashing 68
  installation of equipment 71, *73, 74*
  insulation/ply layout over concrete
      deck *81*
  interior roof drains 69
  locations 66
  parapets/perimeter walls 74–5, *76*
  pitch pockets 67
  reglets 67–8
  relief joint *78*
  roofing man hatches 74
  sanitary vent pipe detail 69–71
  single ply roofing 68
  Standards 82
  thermal expansion/contraction of
      metal 77
  use of aluminum pigmented asphalt
      paint 75
  use of propriety systems 69
  use of scuppers 75
freezer buildings (case study) 144–5

glass fiber felt (case studies) 126–7,
    136–7
global warming 96–7
green construction 95–6

hail storms (case study) 186–7
heavy glass fiber shingles (case
    study) 132–3
HVAC (heating, ventilating, air
    conditioning) 71, 173
hybrid roofs 30
Hypalon polymer 33

insulation 7
  case studies 134–5, 138–9, 140–1, 169
  thermal 43–50

leaks (case studies) 110, 144–5
  background 101
  condominiums 128–9
  distant expansion joints 173–5
  fastener blackouts 152–3
  general observations 101–2
  gymnasium/cast-in-place concrete
      decks 124–5
  lightweight insulating concrete 138–9
  liquid applied waterproofing 167–8
  observations during sampling 102–4
  parapets 117–18
  roof composition 102
  sea gulls 171–2
  two ply problem 157–8
low-slope systems 5
  asphalt membranes 21–4
  average/minimum service life *11*
  building occupancy 13–15
  drainage 19–21
  durability/climate 10–13
  general information 10
  materials used on 5–7, 8–9
  membrane physical properties 15,
      16, 17, *18*, 19
  metal roofing 34–5
  selection avoidance/procedure 14
  single-ply 30–5
  specifications 14
  temperature graph *11*
  use/abuse of building 13

maintenance man (case study) 115–16
materials 5–9
  energy-to-peak strength 17
  mass/equilibrium moisture
      content 17, *18*, 19
  physical properties 15–19
  tensile strength 15, *16*, 17
  see also named materials e.g. APP
      (atactic polypropylene)
membranes
  asphalt 21–4
  asphalt-glass fiber built-up 27–8
  physical properties 15–19
  split (case studies) 107–8, 111–12,
      134–5
metal roofing *see* architectural metal
    systems

National Roofing Contractors
    Association (NRCA) 5
neoprene (chlorinated rubber) 33, 61
noisy roof (case study) 184–5

performance specifications
  latest buzz words 94, 95–7
  standards/test methods 94–5
perlite 7, 48, 49
PIB (butyl rubber) 33–4
plywood, fire retardant (case study) 163–4
polyisocyanurate foam 7
polymer modified asphalt (PMA) 165
polymer modified bitumens 28–30
polymer modified coal tar 30
pressure sensitive products (case study) 161–2
propriety products (case study) 109–10
PUF (polyurethane foam) 34
PVC (poly vinyl chloride) 7, 31–2, 44

roof systems
  buy competence not price 190
  distribution 5
  lessons learned 188–90
  monitor work 189
  overcoming stupidity/ignorance 190
  peer review 188
  sales 5
  types 5
  use ones with successful track record 189
Roofing Consultants Institute (RCI) 158
rust 130–1

SBS (styrene-butadiene-styrene block copolymer) 6, 29, 30
sea gulls (case study) 171–2
SEBS (styrene-ethylene-butadiene-styrene) 29–30
shingles/shakes *see* asphalt shingles
single-ply systems
  attachment methods 35–6
  low-slope systems/materials 30–4
slate 61–2, 63–4
  fake/artificial (case study) 105–6
Standards
  Asphalt Used in Roofing (ASTM D312) 22
  Class A fire resistant shingles (ASTM D3018) 94
  Dynamic puncture resistance (ASTM D5635) 94
  Fastener rupture and tear resistance of roofing (ASTM D4932) 94
  flashing 82
  Hydrostatic rupture resistance of waterproofing (ASTM D5385) 94

Impact resistance of roofing systems (ASTM D3746) 94
slate (C 407) 62
Specification for Cellular Glass Thermal Insulation (ASTM C552) 45
Specification for Cellulosic Fiber Insulating Board (ASTM C208) 49
Specification for Faced Rigid Polyisocyanurate Thermal Insulation Board (ASTM C1289) 48
Specification for Mineral Aggregate Used on Built-Up Roofs (ASTM D1863) 27
Specification for Mineral Fiber Roof Insulation Board (ASTM C726) 47
Specification for Perlite Thermal Insulation Board (ASTM C728) 48
Specification for Preformed, Cellular Polystyrene Thermal Insulation (ASTM C578) 47
Static puncture resistance (ASTM D5602) 94
Tearing resistance of roofing and waterproofing (ASTM D5601) 94
Tensile-tear resistance of roofing systems (ASTM D4073) 94
Test Method for Apparent Viscosity (Flow) of Roofing Bitumens Using Parallel Plate Plastometer (ASTM D4989) 23
Test Method for Flash and Fire Points by Cleveland Open Cup (ASTM D92) 22
Test Method for Softening Point of Bitumen (Ring-and-Ball Apparatus) (ASTM D36) 22
Test Method for Viscosity Determinations of Unfilled Asphalts Using Brookfield Thermosel Apparatus (ASTM D4402) 22
Wind resistance of shingles (ASTM D3161) 94
steel deck corrosion (case study) 130–1
steep-slope systems 5
  architectural metal systems 59–60
  asphalt shingles 52–9
  ceramic tile 63
  concrete tile 63

fiber-cement tiles/slates 63–4
fire classification 52, 53, 57–8
life cycle costs 52, 53
materials used on 7–8
natural slate 612
origins 52
service life 52, 53
structural sheet metal 60–1
wood shingles/shakes 63
structural decks 38–9
   case studies 124–5, 130–1, 136–7
   concrete-excelsior-planks 42
   foamed concrete 41
   lightweight insulating concrete 41
   lightweight structural concrete 40
   lignin-excelsior planks 42
   metal banded gypsum planks 42
   plywood/oriented strand board 42
   poured-in-place concrete 40
   poured-in-place gypsum 41
   precast/pre-stress concrete panels 42
   steel 39–40
   wood fiber plank 43
   wood plank 42–3
sustainable construction 95

tar 23
   see also asphalt and coal-tar pitch
thermal insulation
   case study 182–3
   cellular glass 45

composite 45, 47
expanded polystyrene 47
extruded polystyrene 47
functions 43–5
glass fiber board 47–8
moisture 45
perlite board 48
perlite filled asphalt 49
phenolic foam 48
physical properties 46
polyisocyanurate foam 48–9
urethane foam board 49
thermal splitting 56–7
TPO (thermoplastic polyolefin) 7,
   32–3, 96
trade names 191–234
two ply syndrome (case study)
   157–8

ventilation (case study) 180–1

walkways/roadways 61
waterproofing (case studies)
   improper 178–9
   liquid applied 167–8
websites 235–8
wood fiber/board 7
wood shingles/shakes 63

XPS (extruded polystyrene) 47